Earthbound

OTHER BOOKS OF INTEREST TO EARTH SCIENTISTS

The Third Planet, an Invitation to Geology, Konrad B. Krauskopf

Debate about the Earth, H. Takeuchi, S. Uyeda, H. Kanamori

Affluence in Jeopardy, Charles F. Park, Jr.

The Fabric of Geology, C.C. Albritton, Jr., *Ed.*

Heart of the Earth, O.M. Phillips

Models in Paleobiology, Thomas J.M. Schopf, *Ed.*

Physical Processes in Geology, Arvid M. Johnson

Solutions, Minerals, and Equilibria, Robert M. Garrels and Charles L. Christ

Thermodynamics for Geologists, Raymond Kern and Alain Weisbrod

Theoretical Activity Diagrams, Harold C. Helgeson, T.H. Brown, and R.H. Leeper

Adaptation and Diversity, Natural History and the Mathematics of Evolution, Egbert Giles Leigh, Jr.

Earthbound

MINERALS, ENERGY, AND
MAN'S FUTURE

~~~~~~

Charles F. Park, Jr.

Professor of Geology and of Mineral Engineering

Stanford University

in collaboration with

Margaret C. Freeman

FREEMAN, COOPER AND COMPANY

1736 Stockton Street

San Francisco, California 94133

Contents

Preface

~~~~~~

When *Affluence in Jeopardy* was published in 1968, the author offered it with trepidation as well as with urgency and conviction. *Earthbound* is based on *Affluence in Jeopardy*, although it is a new book, different in emphasis, in scope, and in much of the content. Perhaps the greatest difference is that now, in offering this book, the author feels no trepidation. Time has made real the problems foreseen by *Affluence*—shortages of energy and other mineral commodities, bringing about increasing economic and political turmoil. Now a person can hardly pick up a newspaper without reading of the energy crisis and the growing worldwide competition for all raw materials. In fact, one mineral authority said that the present book should be entitled "We Told You So."

Environmental laws are having an impact, especially in the United States. Construction of the Alaskan pipeline was indefinitely delayed, and drilling for offshore oil and gas has been impeded. About half of the national zinc capacity has been closed, and several copper smelters are barely operative. Wilderness areas are being expanded. The search for new resources is becoming more costly and difficult.

Earthbound explains why civilization must have energy and raw materials as well as adequate unpolluted space, and points out the steps that must be taken to assure their availability. Although the trepidation that marked the offering of *Affluence* is absent, the author's belief that the problems are urgent and his conviction that people everywhere should be informed of them are, if possible, even greater.

Thanks are due to colleagues at Stanford University, especially Evan Just, K.H. Crandall, and F.C. Kruger, with whom much of the subject matter has been discussed. Miss Hazel Blair and Mrs. Eula Park contributed valuable help in reading the manuscript.

We owe an unrepayable debt to Professor Cordell Durrell of the

University of California at Davis, who not only reviewed the material thoroughly and made suggestions for improvement but offered us his independent ideas.

We also thank, for their kindness in permitting republication, Edgar Dean Mitchell, the *San Francisco Chronicle*, the *San Francisco Examiner*, Marshall Ayres, *Bolsa Review*, the United States Steel Corporation, Hollis M. Dole, McGraw-Hill, Handy & Harman, Exxon Corporation, the *Oil and Gas Journal*, The Chase Manhattan Bank, Union Oil Company, Shell Oil Company, Standard Oil Company of California, the American Iron and Steel Institute, and the United States Bureau of Mines.

Charles F. Park, Jr.

Stanford, California
Summer, 1974

Earthbound

MINERALS, ENERGY, AND
MAN'S FUTURE

TO THE READER

This book is written, not for the professional, but for the student or citizen who has very little knowledge about minerals and energy and their relationship to the political economy. We are dealing here with elemental facts at an elementary level, but by the time he finishes the book the reader should see the complexities of civilization's most profound problem: the availability of enough of the earth's finite supplies of materials and energy to assure the continuance of man's life.

1

Man and the Earth

~~~~~~

It's so incredibly impressive when you look back at our planet from out there in space and you realize so forcibly that it's a closed system—that we don't have any unlimited resources, that there's only so much air and so much water.

You get out there in space and you say to yourself, "That's home. That's the only home we have, and the only home we're going to have for a long time."

EDGAR DEAN MITCHELL, *Apollo 14 astronaut*

Right up to the time when the first astronauts went into outer space most people thought of the earth as large and limitless. Through centuries men had explored it. Less than five hundred years ago it provided new worlds to be discovered and many people believed that it extended to infinity.

Gradually, as the human population increased and the earth's boundaries became known, our planet began to seem smaller to its inhabitants, but only when the astronauts actually went beyond those boundaries and took pictures of the earth as a whole could people suddenly view their home in perspective. They saw with their own eyes that its area is limited. They realized that the resources of that area must also be limited.

That marked the beginning of a new way of thinking about the earth. Seeing it as small and limited, most people began to understand what a few had always appreciated—how valuable it is. All of a sudden there was growing concern about the human relationship to the earth and about the effect of increasing human demands upon the earth's finite supplies of what man needs—space, minerals, and energy.

Space

Here we are speaking not of outer space but of space on earth—room in which to live and breathe and move, in which to raise our food supplies, to do productive work, to dispose of our wastes, to maintain some standard of living, and to find something worth living for.

The earth is approximately 7900 miles in diameter at the equator. About 80% of its surface is under water. Of the land, when the permanently arctic and extremely arid regions, those that are uninhabitable or nonarable, are subtracted, there remain about 7.86 billion acres useful to man. This amounts to an average of 2.4 acres for every person alive.

The earth is approximately 4½ billion years old. Man has lived on it for two million years, more or less. Anthropologists estimate the human population of the earth in the Stone Age—400,000 years ago—at approximately 50,000 people (*17*)*. They estimate the world population at the start of the Bronze Age, about 3500 B.C., at approximately five million. At subsequent periods world population was as shown below (*67*).

Year	Millions
1600	450
1700	590
1800	900
1900	1,550

From about 1½ billion in 1900, the world population had grown to 3¼ billion by 1966. The projected rate of growth as estimated by the United Nations for the world is 2.1 percent annually. This means that by the year 2000 the population of the world could be twice what it was in 1966, or about 6½ billion, and the available useful space per person would be down to about 1.2 acres. Each person would have little more than one acre for all purposes—dwelling, crops, transportation, recreation. At this rate, in the subsequent fifteen years (by 2015) the average would be less than one acre, and in the next 7½ years less than half an acre.

Is there any possibility that man, as his numbers increase, may in the future go to other planets that will provide the space he needs? Hardin (*29*) pointed out over a decade ago that if the national standard of living were cut to one fifth its then current level the United States might, using the remaining four fifths of its productive capacity, finance the exportation in one *year's* time of one *day's* increase in the world population to another planet. Of course we would have to

* Numbers refer to the numbered entries of the bibliography, pages 265–268.

find a suitable one, and any search would seem to take us into the realm of science fiction rather than to provide a feasible possibility. The nearest might be a planet of the star Alpha Centauri, which would mean that those making the journey, should technology make it possible for them to travel at seven million miles an hour, would spend about four centuries en route.

Fig. 1-1. The world's population growth. Source: United Nations.

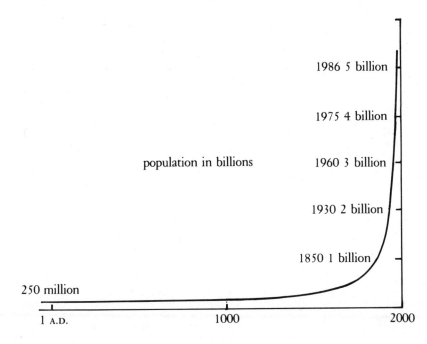

If the earth is the only home we have, how long can man continue to increase when his home is not expandable?

Minerals

Minerals are defined technically as naturally occurring crystalline solids of inorganic origin with a more or less fixed chemical composition. A few substances closely related but not conforming to that definition, such as petroleum, are usually included. In this sense there are about two thousand minerals.

We are interested in them as nonrenewable resources. For our purposes a mineral is defined as a chemical element or compound of chemical elements that exists in nature—that is, minerals are the inanimate substances of the earth. We include, although they are produced by organic processes, fossil and mineral fuels—petroleum, natural gas, and coal. We include soils, but not the atmosphere except in reference to pollution, and we include water only in relation to soils. Sometimes we call minerals "raw materials," for they are the basic substances of industry from the largest organization down to the one-man operation.

Even under our broadened definition there are only about one hundred minerals of economic importance; the table on pages 43–46 lists some that are most useful. Inorganic minerals generally are classified as metallic and nonmetallic or industrial minerals. The metallics include such widely divergent substances as gold, iron, and uranium. The nonmetallics are even more diverse. For example, there are the clays used in the ceramics industry; cement, gypsum, and many other building and insulating materials; the phosphates and other chemical fertilizers; common salt; potassium; sulfur; and the precious and semiprecious stones.

Our need for minerals is as great as our need for space. We need them for food because plants and trees must have mineral nutrients if they are to bear crops. We need them for homes because they provide among other things paint, nails, pipes, wires, and cement. We need them for the manufacture of artifacts, for transportation and communication, as conductors of light and heat, and for the energy required by all activities, human or mechanical.

There is only so much mineral wealth on and in the earth for the use of mankind—and mankind will use it faster and faster as world population grows and as world standards of living improve. The consumption of iron in the world from A.D. 1000 projected to 2000 is pictured in Fig. 1-2. In 1965 the per capita consumption in the United States was about one ton* per year; a comparable world figure was about $\frac{1}{6}$ of a ton per person. To double the population of the world by the year 2000 and simply to maintain the same per capita consumption of iron would mean doubling production, i.e., producing 500 million more tons annually. Should the population of the world as of 1965 raise its consumption, between that date and 2000, to equal that of the United States in 1965, then three and a quarter billion tons, six times our 1965 production, would be required annually. In the event that the population of $6\frac{1}{2}$ billion people in the year 2000 would require one ton per person per year, as did the United States in 1965, then the 1965 annual production of 500 million tons would have to be increased twelve times.

These figures allow for no greater per capita use of iron and its product, steel, than that of the United States, and yet even here people are striving for

---

* In this book, unless otherwise specified, a metric ton is used, equivalent to 2,204.6 pounds. In the United States (the only industrial country in the world where the metric system has not yet been adopted) we use a short ton, weighing 2,000 pounds, and a long ton, weighing 2,200 pounds.

EARTHLINGS

By David Perlman
Science Correspondent

Edgar Dean Mitchell, a pleasant, bearded and immensely skilled young Navy Captain, has come back from the moon with a deep concern for the future of his home planet.

Looking down upon earth during the lunar flight, Mitchell gained a sense of urgency about man's deteriorating environment; he reflected on the destruction of finite natural resources, and he decided to become—as he puts it—"a crusader using reason, not emotion."

At 40, Mitchell is hardly the prototype of the hot test pilot, the space jockey, the gung-ho flyboy. His manner is easy, his speech is quiet, his reflective voice bespeaks conviction and commitment. He is highly articulate and his scientific knowledge is precise: a measure perhaps, of his Ph.D. degree in astronautics and aeronautics from the Massachusetts Institute of Technology.

"It's so incredibly impressive," he said yesterday, "when you look back at our planet from out there in space and you realize so forcibly that it's a closed system—that we don't have any unlimited resources, that there's only so much air and so much water.

"You get out there in space and you say to yourself, 'That's home. That's the only home we have, and the only home we're going to have for a long time.'

"And then you get upset to think of the wars going on down there, of people starving, and of people's lives made miserable and uncomfortable and sick because they can't cope with the way the world has evolved.

"I wish to God I had some pat answers, but I don't."

Mitchell may have no answers to specific problems, but he is convinced that answers can come from science and technology benignly used. He believes that the Space Program can play a major role . . .

Not, he feels, because exploring the moon will solve the problems, but because space exploration can offer basic answers about the history of the earth and the solar system, about energy cycles and geochemistry—and because the space program can mobilize a vast technological productivity and keep it mobilized.

Technological solutions to the environment problem must be found and applied quickly to industry—through voluntary action where possible, and through uniform regulation where industry refuses to see its own "self-interest," Mitchell believes.

To Mitchell the problems of energy and over-population are the critical ones. Societies are growing, expanding, industrializing. Power consumption is rising.

"So we simply must find alternate energy like nuclear power," Mitchell said. "But even this is only a stopgap until we discover the ultimate, constant, undiminishing energy source."

From *The San Francisco Chronicle*.

Fig. 1-2. World consumption of iron ore. Source: United States Bureau of Mines.

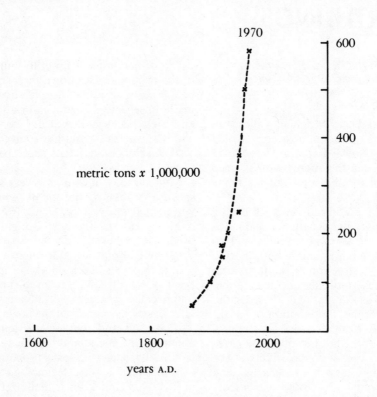

higher standards of living that will require greater per capita consumption of both.

The rate of consumption of two other commodities, copper and lead, is shown in Fig. 1-3. To raise worldwide standards of living in the year 2000 to those of the United States at present would require the staggering annual production figures of 53.3 million tons of copper, or approximately eleven times the amount now being produced, and 35.5 million tons of lead, or sixteen times the present output.

People are often misled by statements in the newspapers or on the television giving estimates of the total resources of a raw material Resources are not reserves. Resources are what exists; reserves are what you can get out. For example, the National Petroleum Council issued in December of 1972 a summary of oil shale resources in the Piceance Basin, Colorado.

1. Total estimated resources .... 1,200 billion barrels
2. Total resources in beds more than 30 feet thick,
   averaging more than 30 gallons per ton .... 120 billion barrels
3. Estimated recoverable reserves .... 50 billion barrels

Fig. 1-3. Copper and lead: world per capita consumption, 1960 vs. 1970. Source: United States Bureau of Mines.

World 1960 population: 3 billion people
1970 population: 3.3 billion people
U.S.A. 1960 population: 180 million people
1970 population: 200 million people

Copper consumption
World 1960: 4,724,000 short tons = 3.15 pounds/person
1970: 7,139,000 short tons = 4.32 pounds/person
U.S.A. 1960: 1,422,000 short tons = 15.0 pounds/person
1970: 2,079,000 short tons = 20.79 pounds/person

Lead consumption
World 1960: 2,629,000 short tons = 1.75 pounds/person
1970: 3,707,000 short tons = 2.25 pounds/person
U.S.A. 1960: 1,021,000 short tons = 11.4 pounds/person
1970: 1,422,000 short tons = 14.22 pounds/person

On a worldwide basis, including the U.S.A., consumption over the period 1960–1970 has increased:
Copper: 1.17 pounds/person in world, or 37%
5.79 pounds/person in U.S.A., or 39%
Lead: 0.50 pounds/person in world, or 29%
2.8 pounds/person in U.S.A., or 25%

Thus a resource of 1,200 billion barrels means a reserve of 50 billion barrels.

The fact that resources are not reserves is true of all nonrenewable resources. There is copper in granite, or iron in your own back yard. Each is a resource, but not a reserve under any conditions you can think of.

Dole (*13*) recently pointed out that the United States alone has consumed an estimated 260 billion dollars' worth of nonrenewable raw materials since 1940. Prior to that date, total world consumption reached only 250 billion dollars, based upon 1929 prices. Mineral consumption is increasing throughout the rest of the world more rapidly than in the United States, and this is bringing about ever increasing competition for raw materials and energy in the world markets (*68*).

The earth's ore bodies, which formed at irregular intervals during the 4½ billion years of geologic history, are exhaustible. If we continue to consume them until the remaining mineral supplies are not adequate for our needs, how will we get along without them?

Fig. 1-4. Increase in world consumption of copper during the last century and a half. Source: United States Bureau of Mines.

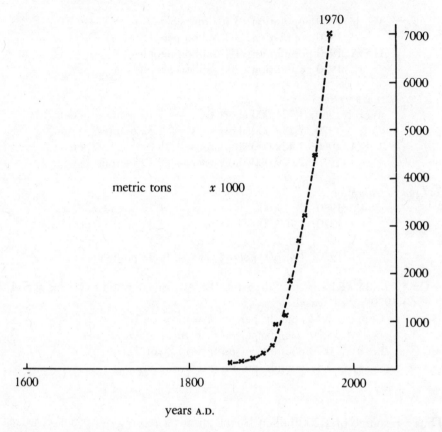

Substitutes. Surprisingly large numbers of educated, but unknowledgeable, people are able to dismiss that question with the answer that technology will provide substitutes. At present, substitutes are available for many uses of many minerals, but in some important areas no substitutes have yet been found. Nothing now known can take the place of steel where strength is needed, as in skyscrapers and dams or in the high-temperature alloys for parts of a jet engine; nothing now known will substitute for cobalt in the manufacture of the strong permanent magnets needed in all modern communication systems; and no other metal will, like mercury, remain liquid at ordinary temperatures so that it is usable in temperature and pressure control equipment. Substitutes have not been found although the twentieth century has seen great amounts of time, energy, and ingenuity go into the study of possible combinations of the elements. There are in nature only ninety-two chemical elements; the number of materials that these in combination can make is limited. And where can any substitutes come from? We have only the earth.

Fig. 1-5. Increase in world consumption of lead during the last century. Source: United States Bureau of Mines.

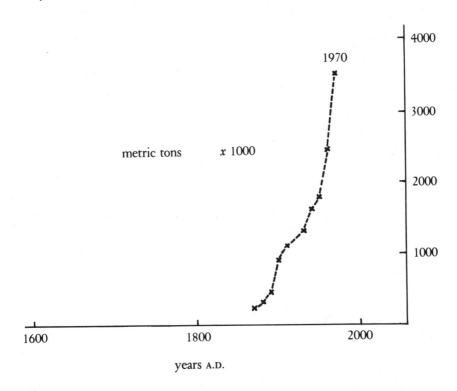

metric tons   *x* 1000

years A.D.

Ocean mining. There is hope that at some time in the future we may turn to the oceans for many of the mineral resources we need. At the moment this is not economically feasible except in shallow waters; the mining of deep-sea deposits presents numerous complicated problems of technology and jurisdiction because of their location. Moreover, the extraction of minerals from the ocean will require abundant inexpensive energy. And are the oceans inexhaustible?

Fig. 1-6. The United States used in 1971 the following percentages of world production of six major resources. Source: United States Bureau of Mines.

Petroleum	. . . . .	32%	Steel	. . . . . .	19%
Natural gas	. . . . .	57%	Aluminum	. . . . .	35%
Coal	. . . . . .	16%	Copper	. . . . .	27%

Recycling. Recycling is a necessary means of conservation of our mineral resources. In many crucial cases, however, the reclamation of minerals is uneconomic. For example, the two largest uses of silver today are in photography and electronics. Great as the sum total of these uses is, the amount used in each unit is so small that one cannot imagine the economic recovery of silver in sizeable amounts from photographic film used by individuals or from outmoded electronic devices. Similarly, it is difficult to conceive of the reclamation of mercury from broken thermometers.

In addition all recycling requires energy, including the costly muscle energy of collecting. The Sanitation Workers' Union in New York City in 1970 computed the annual cost of collecting discarded newspapers (about four or five hundred thousand tons of them) at $13,260,000, which amounts to approximately $30 a ton. The rest of the recycling process costs far more than the collecting. The result of this recycling is paper more costly than new paper, and inferior in quality.

When we speak of the cost in money, the money is a symbol representing the cost of the energy that has to be expended in recycling, mining, or any other activity.

Energy

Energy may be defined as *the capacity to take action* (although a physicist would insist upon a more precise definition).

People require energy, as they do space and minerals, merely to exist. Since food as a topic does not come into the scope of our discussion, here we consider the important requirement of food as implicit in the need for energy. Food is a form of energy that can be provided for animal life only by plant life, using water, mineral nutrients, and sunlight.

Any living organism requires energy to live—to grow, to sleep, to breathe, to work, or to play. All transportation necessitates energy. All minerals in the ground are useless without it, for energy is needed to extract them. It takes energy, human or mechanical, to make or build anything, from the simplest item to the most complex. Energy is so essential that the quality of our life depends on the amount of clean, inexpensive energy available to each person.

There are many sources of energy, but at present man has not learned how to utilize any source that can provide energy in unlimited amounts. Energy from human muscle is the most limited, for it produces the least results. Petroleum has been the most widely used, because it has been cheap and most easily transported and handled, but it is exhaustible and causes pollution of the environment. Natural gas, the cleanest mineral fuel, is cheap, but known reserves of it are decreasing in the United States. Coal is cheap, but it is dirty and expensive to transport; it, too, is exhaustible. Hydroelectricity is limited, for in industrial regions most of the sources of power from falling water have now been harnessed. Nuclear power accounted for only two tenths of one percent of the total energy used in the United States in 1970, after twenty-five years of attempts to utilize it properly. The major problems are the disposal of its dangerous radioactive waste products and the fear that people have of leakage of radioactivity. Also, it causes pollution through the effect of its surplus heat upon the environment.

Only solar energy, hydrogen, the tides, and the heat radiated from the earth's interior seem inexhaustible, and we have not yet succeeded in harnessing effectively these potential sources.

The energy crisis so vital today means more than an acute shortage of gasoline and fuel oil. It means the prospect of not having the energy necessary to produce any of the basic raw materials and the processed goods that come from them.

How is man to continue to get, in increasingly large amounts, the cheap, clean energy he needs?

Complexities

The problem appears difficult enough: Man's demands upon the earth for his basic needs of space, minerals, and energy are increasing, while the earth remains a closed system, with nonrenewable resources. But in addition there are complications:

1. Man's three demands are interrelated—each has a physical effect upon the other two. Probably no reader would approve of what strip mining has done to the mountains of West Virginia. At the other extreme, is the mining of needed minerals to be reduced because environmentalists impose such stringent restrictions upon operations that the owners of many mines cannot afford to stay in business? We wish to conserve raw materials by recycling. But recycling on a large scale requires energy and other raw materials for gathering, sorting, and transportation, and for the manufacture and operation of heavy machinery. Some of these processes would add to our pollution problems. We wish to increase our use of such substitutes as plastics, to save minerals, but plastics require minerals and energy for their manufacture, and create further problems of disposal. Electricity, gasoline, and minerals in general must be transported in order to be available everywhere they are needed. Yet whether the means is by power lines or by trucks over a highway system, space and minerals are required and energy is consumed, and some modes of transport pollute the air.

2. Each of man's three demands has a strong influence upon and is influenced by economics and politics.

No industrial nation is self-sufficient in all the minerals it needs. Every nation depends upon others for at least part of its raw materials. Hence there arises the necessity to trade; countries must buy and, in order to have the money to buy, they must sell. The United States, for example, has within its boundaries adequate supplies of less than a dozen of the hundred minerals it requires; the others are imported in part or in whole.

The impact of a nonrenewable resource on politics and economics is probably best illustrated by petroleum. Oil is the most important single commodity on the international trade list, from the standpoint of both volume and value. Every country is involved in the petroleum and petroleum products trade, either as a producer or as a consumer. Because of the great quantities sold and the universal need for products manufactured from it, petroleum is a peculiarly convenient target for all kinds of projects for raising money and improving the financial status of one country at the expense of another. And it should not be forgotten that the inexpensive and abundant energy available from petroleum is a basis of modern industrial civilization everywhere, while about two thirds of the known reserves are in one area, the Middle East.

The fact that competitive sources of petroleum were available in the past served to stabilize the market and prevent prices from getting out of hand. At present the likely possibility of future shortages in the industrial nations is reacting to the benefit of the producing countries, since it seems certain that everyone will pay more for energy used in the future. If, however, the increase in cost is so large that consumption must be drastically curtailed, then the standard of living of people in both the consuming and producing countries will go down.

Economics and politics are affected by man's increasing need for space as well as by his increasing demands upon finite supplies of minerals and energy. The rapidly spreading pollution of the environment has become a problem of growing concern all over the world. Governments have established environmental control agencies. Industries are no longer permitted to pollute the air and water. Regulations are more and more extending to the prohibition of pollution by individuals. Yet pollution is an international problem while control is to date limited to local actions.

No one can deny the necessity to protect the environment, but if in the United States the recent regulations on the use of smelters force many mines to close, how many jobs will be lost? What will be the effect upon the nation's labor force and thus upon its economy, its Gross National Product, and its balance of trade?

Population control is now seen by most thinking people as a necessary means of providing for sufficient space in the future. Yet there are nations in Africa afraid to limit their populations because their neighbors may not do the same.

Fig. 1-7.   The United States imported in 1971 the following percentages of the minerals it needed. Source: United States Bureau of Mines.

Rutile	100%	Gold	48%
Asbestos, long fiber	100%	Bismuth	47%
Mica, sheet	100%	Tantalum	46%
Strontium	100%	Potassium	45%
Manganese	96%	Zinc	45%
Aluminum	92%	Mercury	42%
Chromium	89%	Silver	34%
Fluorine	80%	Iron	30%
Asbestos, common	76%	Vanadium	28%
Cobalt	75%	Petroleum	24%
Platinum group metals	75%	Lead	18%
Nickel	66%	Copper	6%
Tin	64%	Natural Gas	4%
Antimony	54%		

Less than 1%--Coal, boron, diatomite, lime, magnesium metal, molybdenum, phosphorus, sand and gravel, sulfur, tungsten, and uranium.

And in 1972 the rulers of Russia encouraged the country's citizens to have larger families. A little earlier the president of Mexico, although he later reversed his position, proclaimed that "greater numbers mean greater power."

The growing necessity for international cooperation is obvious. As the earth becomes more crowded, as demands upon nonrenewable resources increase, what will be the future of earthbound man if he does not have an intelligent plan that provides for the needs of all nations for space, minerals, and energy?

What we can do

Young people predominate among those who are aware that mankind is facing a problem. More than their elders, in general, they are concerned about the environment and the conservation of nonrenewable resources. They may advocate Population Zero, or take newspapers, bottles, and tin cans to recycling centers, or use bicycles instead of automobiles, or carry groceries in string bags to save paper and thus trees. Their concern is more than commendable; it is essential. As it is usually expressed, it holds only one danger.

Because of the interrelation of man's needs for space, minerals, and energy, any effort to influence one of them alone may actually increase the problems of the others. A person who states categorically that every operation causing pollution of the environment must be stopped immediately is not sufficiently well informed; he does not know what the result would be on supplies of minerals and energy, and on the quality of living. An extremist throwing rocks at automobiles while he shouts, "Ecology now!" is so ignorant in his approach to the problem that he can achieve no solution whatever.

To find any solution, to make any intelligent plan, we must study the whole problem of what man requires from the earth—not part of the problem, but all aspects and all their interrelationships. The first thing, therefore, that any concerned individual can do is to acquire knowledge of the earth and its resources of minerals and energy.

In this book, which attempts to provide adequate essential information, the reader will learn about the principal classes of minerals, about energy, and about soils and water. He will then see how economics and politics relate to the extractive industries, and how conservation and the environment affect and are affected by the preceding subjects. Finally, he will be ready to understand what man can do to insure sufficient supplies of nonrenewable raw materials, including energy, and to improve the quality of living.

Before we go on to those topics, we deal with what might be called the intermediaries between the citizen on one hand and minerals and energy on the other, since the citizen, in spite of their vital importance to him, has no direct experience in obtaining energy or minerals. They are provided for him by the extractive industries.

2

The Extractive Industries

~~~~~~

The extractive (or minerals) industries are those that explore for raw materials and remove them from the earth. They may be grouped into three general categories of activities: the mining of metals; the extraction of nonmetals, such as salt and sulfur, sand and gravel; the mining or drilling for energy materials—coal or oil or gas.

These industries are in the main not well known to the public because most of them have not heretofore been primarily retailers. Their major function has always been to supply to fabricators nonmetallic minerals or roughly refined metals—sand and gravel to road builders; gypsum to manufacturers of wallboard for housing; copper, iron, lead, and other metals to various businesses for use in the manufacture of automobiles, telephones, refrigerators, and numerous other items familiar in everyday living. The energy industries, which supply gasoline to the consumer as well as coal, gas, and fuel oil to public utility companies, are distributive as well as extractive, and thus closer to the usual retail business. All of the extractive industries are the suppliers of the fundamental materials that form the bases of any nation's industrial structure and economic welfare. If there were a shortage of these raw materials they would have to be rationed to industries, which then could not produce the supplies or the income people want for their standard of living. Instead of industrial expansion, there would be industrial and economic regression. Without these raw materials, there would be neither modern industries nor the products people today demand.

Because they are generally unknown to the public and because of their distinctive characteristics, the minerals and energy industries have not been understood by the average person. Most of us have accepted without question the ready availability of materials: gasoline and lubricants could until recently be purchased everywhere, chromium and nickel are simply part of the stainless steels that we have no trouble finding in stores, tin cans are on any grocery shelf. We give scarcely a thought to where the ingredients came from, or how they have been made available to us. The gasoline could be from Louisiana, Venezuela, or the Middle East; the chromium in the stainless steels is apt to have originated in South Africa, Rhodesia, or Russia; the nickel was mined in Canada; and the tin of the cans probably came from Malaysia or Nigeria. The development of the system that has enabled these and other mineral products to be so widely distributed and so readily available is one of the most fascinating stories of modern times. It is the story of the extractive industries.

All of the extractive industries must deal with three unusual problems:

- Irregular distribution of mineral deposits, including fuels.
- High capital cost of the risk of exploration and development.
- Exhaustibility of the chief asset.

Those industries that extract certain metals have in addition a fourth problem:

- Indestructibility of the product.

No other business faces problems similar to these. They are the major reasons why the extractive industries are unique.

Irregular distribution

Why does the United States have to import all of its manganese, industrial diamonds, chromite, and tin, and most of its fluorspar, tungsten, and other minerals, when territorially it is the fourth largest nation in the world? In the past, its prosperity has in large part depended upon the ready availability of raw materials and cheap energy, but demands for all mineral commodities have grown tremendously in response to spreading industrialization. In spite of extensive search by geologists and geophysicists, using the most modern methods, domestic discoveries are now far too few to enable production to keep up with these demands. Deposits of many materials have not been found or, if found, have been too small, too low in grade, or too difficult to treat metallurgically to permit competition with materials available elsewhere on the world market.

The nations with the greatest need are now lacking in adequate supplies; Europe and Japan, to an even greater extent than the United States, are dependent upon the import of petroleum, copper, and other commodities to sustain their industries and their way of life. In fact, no industrialized nation is today self-sufficient although many nations have tried to attain this status. All must import part of the raw materials they need to maintain their modern industrial civilizations.

The locations of mineral deposits are fixed by nature, which seems to have scattered them haphazardly and whimsically over the world. For example, about two thirds of the world's copper is produced by four countries—Zambia, Chile, Zaire (formerly the Republic of the Congo), and Peru. None is a large user of the metal. They all mine it principally for export and they together dominate the market. Petroleum also is concentrated in occurrence. About two thirds of the world's reserves are in the Middle East, which consumes little but relies heavily on income from the export of its oil.

Mineral deposits have always been prizes that nations struggled to win. From the time when the Romans seized the tin mines of Cornwall and the mercury and gold of Spain, Europe's history has been marked by wars precipitated largely by attempts to control deposits of lead and zinc, iron ore, coal, and other mineral wealth. The Spanish conquest of the Western Hemisphere was based to a considerable extent upon the desire for gold. At the very roots of modern history are efforts to acquire known mineral deposits: the iron ores of Alsace-Lorraine and the coal deposits of the Ruhr where France and Germany meet; the excellent coking coals of Silesia that have made Poland a target for conquest; and the nickel deposits of Petsamo in the northern Kola peninsula which Russia annexed during World War II "in order to protect Murmansk from Finnish aggression."

At the present time, impending mineral shortages, caused by greatly increased international demands, are casting their shadows over many nations. Producers are raising prices, governments are expropriating private property where they consider it advantageous, and diplomatic problems are growing for the future. Competition to obtain needed resources is rapidly approaching a point where some of the poorer nations will be unable to get the raw materials needed for an improved standard of living.

The high-grade mineral commodities that are localized in only a few countries or districts are particularly susceptible to pressures and controls by governments and by cartels and monopolies, much more so than widely distributed commodities. Governments can exert pressures by limiting the amounts of exports and imports, levying taxes, fixing prices, setting depletion allowances, stockpiling, subsidizing, selling government-controlled stocks, and, as a final constraint, by expropriation and government operation.

Cartels and monopolies operate most effectively when more than 70% of a commodity is under close control. The mining and marketing of diamonds has been for years a monopoly of the DeBeers Syndicate, which has been offered serious competition only with the development of the processes for manufacturing commeercial-grade diamonds for industrial uses.

Another example of a monopoly made possible by a small number of producers is the International Tin Committee, which includes representatives

Fig. 2-1. World consumption of certain metals, 1960–1970. Source: *Bolsa Review*, Vol. 5, No. 56, p. 456, August 1971.

| | *Thousand tons* | | *Percentage* | *Annual percentage* |
	1960	*1970*	*increase*	*growth rate*
Aluminum	4,177	9,739	133%	8.8%
Copper	4,724	7,139	51%	4.2%
Lead	2,629	3,707	41%	3.5%
Nickel	293	558	90%	6.6%
Tin	201	240	19%	1.8%
Zinc	3,088	4,859	57%	4.6%

from both producers and consumers. In trying to stabilize the price of tin, the Committee has set a floor and a ceiling; if the market price goes below the floor, the Committee buys tin, which it will hold in stock. If the market goes above the ceiling price, the Committee will sell the metal until the price is again reduced. In general this Committee has aided considerably the fiscal planning of the producing nations by stabilizing prices.

The irregular distribution of mineral deposits means that the industrialized nations must import the raw materials necessary to sustain their industries. Japan must bring in most of what it needs—iron and coking coals, for example. The United States must import, in varying percentages of the amounts used, eighty-eight or more of the hundred mineral products most essential to the nation's industrial complex.

Many of the raw materials, such as copper and petroleum, needed by the industrialized nations are obtainable only from the underdeveloped countries. Raw materials thus furnish a major part of the support available to many developing nations, all of which are anxious to get better prices in order to increase revenues. The irregular distribution of mineral deposits means that developing nations with surplus commodities may export. It also means that many underdeveloped nations are overly dependent upon income from the sale of a few commodities or even a single one. Venezuela depends upon oil and iron, Chile and Zambia upon copper, the countries of the Middle East upon oil, Bolivia and Malaysia upon tin; the list is long.

Irregular distribution means that extensive international trade must be carried on safely, rapidly, and freely, which at present is not always possible because, in addition to numerous government restrictions, international commerce calls for shipping arrangements, marine insurance, tariff and customs procedures, duties, monetary exchanges, and scores of other regulations.

Many minerals are restricted, not only in their distribution, but also in the amounts available in the known deposits. For example, in spite of the fact, recognized for years, that beryllium when added in very small amounts to copper makes a strong, hard, and superior metal, there has been insufficient beryllium to allow more than a limited use of it as an alloy. Only within the past few years have new discoveries made larger amounts of beryllium available. Silver and gold and many other metals are still available only in restricted amounts; large new deposits have not been found in spite of extensive exploration.

High risk of exploration and development

The uncertainty of discovery during the search for deposits introduces into the minerals industries a high degree of risk as well as the possibility of speculative gain. Nearly everywhere in the world, the fields of common minerals that are exposed at the earth's surface have already been found. Extractive companies no longer have much fear of competition resulting from new discoveries and oversupplies now that the demands for all mineral products are so great. Rather, the emphasis is on the need to find new deposits and on the question: How can more minerals be found?

As the search for additional resources is intensified throughout the world, the problems and difficulties of discovering new deposits also increase. Costs have grown tremendously. Many companies now estimate that an expenditure of a minimum of twenty million dollars is the average required to find one deposit worth development—and this cost includes nothing of the many millions required for the thorough final evaluation of the deposit and its preparation for mining. The cost of unsuccessful search for oil and gas is also high; one estimate puts the drilling of dry holes in the United States at a billion dollars a year.

The search for minerals is in fact becoming so expensive that for most individuals and many nations the costs are prohibitive. With few exceptions, only large, well managed companies, with capable exploration and operating staffs and adequate risk capital backing them, are able to survive with profit for more than a few years. Even if an extractive company spends money unstintingly on exploration, no one can know in advance whether or not a search for a mineral deposit will be successful, or how much time and money will be needed if it does eventually succeed.

A geologist may examine several hundred prospects before he finds one he can recommend. A mining executive recently stated that examination of about a thousand prospects was required to find one mine.

Few places in the world have not been examined and re-examined in the search for minerals; only those areas covered by soils with thick vegetation, by snow and ice, or by shifting sands remain unexplored. Even the ocean basins are being scrutinized closely and minerals are being recovered from the shallower depths along the continental margins. Search is being extended into the deeper ocean basins, at depths of fifteen thousand feet or more, and methods of dredging or recovering the nodules found on the ocean floor at those depths are under study.

On the land, lower and lower grades of ore are being mined. Few people would have thought, only a few years ago, that rock containing 0.35% copper (seven pounds per ton of ore) or 25% iron in magnetite would be ore, but they are. The decrease in the grade of the ore being mined means that very large amounts of ore, at places more than 100,000 tons a day, must be handled and the margin of profit per ton is small. The development of large and ingenious devices for dealing with huge tonnages of earth and rock have made this type of mining possible.

The open pits and caved areas that result from large-scale mining operations constantly bear the brunt of the objections of environmentalists. Costly planning to meet environmental requirements may cause a deposit to be uneconomic. Many deposits in the United States that might be contributing to the national wealth are now idle, partly because environmental costs are prohibitive and partly because of political pressures exerted by groups of conservationists. The extraction of raw materials is not permitted in many places and is unwanted in many more. Yet the locations of minerals remain fixed by nature and the deposits cannot be moved.

In underground mines the difficulties of finding additional ore reserves increase as the depths of the mine openings increase. Imagine the costs and problems involved in probing for small ore bodies in rocks a mile or more below the surface of the earth. No wonder the grade of ore mined at depth must be considerably higher than the grade of the materials recovered in surface open pits.

Inexperienced people frequently ask why a grid system should not be laid out in an area where ore is being sought and drill holes sunk at regular intervals on this grid. The cost of such drilling would be prohibitive—and the deeper the drilling, the more costly. Any possible profit from the mining of ore would quickly be consumed by this action. Even in socialist and communist nations where government controls both resources and exploration, and where costs are frequently of secondary consideration, such methods are recognized as both excessively expensive and unproductive.

Modern effective mineral exploration is a team effort that requires the use of many skills; the day of the lone prospector and his burro is long gone. Carefully coordinated and costly geological, geophysical, and geochemical studies combined with specialized aerial photography are the current methods used in the search. First an area to be explored must be selected and the commodity or commodities sought must be designated. No geologist would look in beach sands for deposits of zinc or in granitic rocks for nickel. When the area and the commodities are chosen, geologists study the character of the rocks, their composition and the structure of the earth near by. Geochemists sample the soils, stream sediments, vegetation, or rocks for trace amounts of various elements, and geophysicists try to find areas in the earth showing anomalous physical properties that may indicate the presence of ore. These data are brought together and a great deal of the result depends upon the skill of interpretation.

When a favorable target is finally agreed upon, the area must be explored and sampled. This is generally done with a diamond drill from which a thin core of rock is obtained. The core is usually split, part being sent to a laboratory for analysis and part being filed for reference. Quite commonly two to four years of drilling and trenching are required before a decision can be reached as to the value of the property. If the deposit is too low in grade or too small or too remote, it must be written off as a loss.

The exploration for and the development of ore deposits are among the frontiers of modern science. As problems increase annually, great ingenuity and creativity are needed to furnish man the raw materials he needs to maintain industry and to improve his standard of living.

When a new discovery of any mineral commodity is made it will affect the existing industry according to the grade, the size, and the location of the deposit discovered. For example, the price and profit structure of the lead industry would be altered if a new large deposit of lead, such as that of southeastern Missouri, should be found. This is particularly true if the new deposit is situated near a consuming center where transportation rates are reasonable and total costs can be kept below market prices. The effect is the same as if the United States government were suddenly to release large amounts of lead from the national stockpile at less than the prevailing price. In the base metal industry, a change of one or two cents per pound of metal may determine whether or not the operation is profitable. In the iron and steel industry, where a new furnace may cost tens of millions of dollars, the profit margin is even less than in the base metals, since iron is historically one of the cheapest metals.

The unforeseeable problems that add to the risk and the costs of exploration, development, and mining are many. They may be due to hazardous conditions or difficulties of ventilation and cooling, drainage, or hoisting. In open pit mines, landslides and flooding or other unexpected events may delay production and add to costs. In coal mines, noxious or explosive gases may be encountered. There may be a decrease in the value of any ore at depth, or the ore body, as mining increasingly exposes it, may break up into small units that cannot be handled profitably. At times an increase in the flow of water adds to costs; in extreme cases, it may cause abandonment for a while of even a rich ore body.

Depth in underground mining presents problems not found in open pits and shallow workings. Temperatures of the rocks increase with depth, the average increment being about 1°F for every hundred feet of depth, although wide variations from this average exist. In some mines, a depth is reached where it becomes impractical or even impossible for men to work unless the air is cooled and dehumidified. Air conditioning and ventilation under these conditions may require great ingenuity and certainly necessitate an added investment.

Swelling ground, a phenomenon which is more apt to be met as the depth of mining increases, also increases costs. In several deep mines, such as those of the Champion Reef in the Kolar Gold Fields in India, the weight of the overlying rocks slowly squeezes the walls of the mine into the openings until these are closed. In other deep mines, where the wall rocks are brittle, the pressures are relieved with explosive violence in the dangerous phenomenon known as a rock burst.

The hoisting of ore and waste rock to the surface is still another factor in the greater energy requirements and higher costs of deep mines. Mine shafts are fixed in size, with specific capacities, and the ore hoists operate at set speeds. Hoists operate exactly as do elevators in buildings. To raise a given tonnage of ore takes longer from deeper levels than from those nearer the surface, and in a given period of time less rock can be hoisted through a shaft from deeper than from shallower levels. The capacity of a shaft can be altered only by the installation of faster hoisting equipment or enlargement of the shaft. In contrast with open pit mining, in which output can usually be expanded rapidly, underground mining requires time and large amounts of capital for increased output.

The cost of mining increases also when men must be transported to deeper levels and for longer distances to operating faces underground, since all working places cannot be close to the entrance shafts. Some mine operators find it advantageous to maintain separate shafts for the transportation of men and equipment and for hoisting ore and waste rock.

A problem increasingly encountered by operators of small mines is their inability to find smelters to handle their production. Smelters, some of which are old and not highly profitable, have suffered more and more frequently the attacks of environmentalists; several within the United States have been forced to close and others have had their operations drastically curtailed because their owners believe that modernizing the equipment would not repay its cost.

The many problems and possibilities of discovery and development lend a zest and a challenge as well as a risk to any extractive industry. It is no place for the faint-hearted.

Fig. 2-2. The principal ore minerals.

Products	Mineral Source	Chemical Nature
Aluminum	Boemite (bauxite)	$Al_2O_3 \cdot H_2O$
Antimony	Chiefly a byproduct (see copper)	
	Stibnite	Sb_2S_3
	Tetrahedrite	See copper
Arsenic	Arsenopyrite	$FeAsS$
Asbestos	Chrysotile	$3MgO \cdot 2SiO_2 \cdot 2H_2O$
Barium	Barite	$BaSO_4$
Beryllium	Beryl	$Be_3Al_2(SiO_3)_6$
Bismuth	Bismuthinite	Bi_2S_3
Cadmium	Greenockite	CdS
Calcium	Calcite, abundant in limestones and dolomites	$CaCO_3$
Chromium	Chromite	$FeO \cdot Cr_2O_3$
Cobalt	Smaltite	$CoAs_2$
Copper	Chalcocite	Cu_2S
	Chalcopyrite	$CuFeS_2$
	Bornite	Cu_5FeS_4
	Tetrahedrite	$Cu_{12}Sb_4S_{13}$
Fluorine	Fluorite (fluorspar)	CaF_2
Gold	In its native state	Au
Iridium	A byproduct. One of the platinum metals	Ir
Iron	Hematite	Fe_2O_3
	Magnetite	Fe_3O_4
	Pyrite	FeS_2
Lead	Galena	PbS
Lithium	Found in pegmatites and in brines	Li salts
Magnesium	Recovered from ocean brines	Mg salts
	Dolomite	$(Ca,Mg)CO_3$
Manganese	Pyrolusite	MnO_2
	Psilomelane	$(Ba_3,H_2O)_4Mn_{10}O_{20}$
Mercury	Cinnabar	HgS
Molybdenum	Molybdenite	MoS_2
Nickel	Pentlandite	$(Fe,Ni)S$
	Garnierite—a hydrous nickel magnesium silicate	
Niobium (Columbium)	Columbite	$(Fe,Mn)Nb_2O_6$
Osmium	One of the platinum metals	Os
Palladium	One of the platinum metals	Pd
Phosphorus	As phosphates from sedimentary rocks.	
	Apatite	$Ca_5F(PO_4)$

Fig. 2-2.   (Continued)

Products	Mineral Source	Chemical Nature
Platinum	In its native state with other metals	Pt
Potassium	Sylvite	KCl
Rhodium	One of the platinum metals	Rh
Selenium	A byproduct. Mainly with copper ores	Se
Silver	In its native state	Ag
	Acanthite (argentite)	Ag_2S
Sodium	Halite (common salt)	NaCl
	Trona	$Na_2CO_3 \cdot HNaCO_3 \cdot 2H_2O$
Sulfur	From salt domes Byproduct from petroleum and smelting	S
Tantalum	Tantalite from pegmatite rocks	$(Fe,Mn)Ta_2O_6$
Thorium	Thorite	$ThSiO_4$
	Monazite	$(Ce,La,Y,Th)PO_4$
Tin	Cassiterite	SnO_2
Titanium	Rutile	TiO_2
	Ilmenite	$FeTiO_3$
Tungsten	Scheelite	$CaWO_4$
	Wolframite	$(Fe,Mn)WO_4$
Uranium	Uraninite	UO_2
	Carnotite	$K_2O \cdot 2U_2O \cdot V_2O_5 \cdot H_2O$
Vanadium	Smelter byproduct. Common with uranium	V
Zinc	Sphalerite	ZnS
Zirconium	Zircon	$ZrSiO_4$

Exhaustibility of the chief asset

All deposits of minerals and mineral fuels are finite. Operators in both mines and oil fields are steadily consuming their principal assets, which are irreplaceable. In order to stay in business, companies must amortize their investments, secure their profits, and finance the exploration and development costs of new deposits while the operating mine or oil field is still productive. When a mineral field is finally exhausted (and all are eventually), it is abandoned; there is no turn-in value on a worked-out mine or a dry oil or gas well.

Absolute exhaustion of a mineral property seldom occurs suddenly; there is instead a slow decrease in quality of the deposit and in its value, and an increase in costs of production. The economic limit of a mineral property is usually anticipated well in advance of exhaustion and closure.

In mining, the first blush of discovery of a deposit generally results in the extraction of the best grade of ore, partly in order to amortize the investment as quickly as possible. This is usually followed by a period during which lower and still lower grades of ore are extracted, while technology must be improved in both mining and recovery, and better transportation facilities developed. Eventually, however, a point is reached, even in the best of mines, at which costs have increased and output per man has decreased so that profits vanish. This may result from the need to mine at greater depths, or from any of numerous other causes.

Whatever the cause, costs ultimately exceed the value of the production and a mine becomes uneconomic to operate. The property is then closed unless the government sees fit to subsidize or take over its operation. The abandonment of a mine is usually followed by the forced migration of its employees, which means the loss of personal property, costs of moving and transportation, a search for new jobs, frequently in an industry with which the displaced employees are not familiar, and problems caused by the adjustment of workers and management to a new environment. Therefore, in many countries, though not in the United States, mines that would otherwise close are subsidized, partly for socio-economic or political reasons to prevent large-scale displacement of labor. The operation by subsidy of mines that are marginal or uneconomic for private enterprise means also that valuable resources can be saved for the use of the general economy.

At the end of World War II, when the Boleo copper mine in Baja California, Mexico, became unprofitable to operate, the government took charge. It has operated the mine ever since at little profit but not at a loss. The small town of Santa Rosalia, whose inhabitants depend upon the mine for their livelihoods, has thus been kept alive and the country obtains a considerable amount of copper that it needs and would otherwise have lost. Were it not for government intervention, the Boleo mine would have been deserted, as would also the town of Santa Rosalia, whose people would have had to search for employment in the already overcrowded Mexican labor market. There are many mines that have been maintained by government operation.

Similarly, when oil and gas wells reach a point at which production and transportation costs equal or exceed the value of the product, the wells must be abandoned unless the government intervenes or a new technique is found to permit additional recovery.

Given the fact of exhaustibility, one may wonder how so many mines have continued to be active for very long periods of time. The answer lies in the development of new, advanced mining techniques and the design and manufacture of better machinery and equipment. Improved technology permits the handling of larger tonnages of lower grades of ore at lower costs. A cliché in mining states that the uneconomic minerals of today are the economic ores of tomorrow. For instance, the percentage of copper required for an ore to be economic was in the early 1920's about 1.5%; in 1970 it was about .5%. In other words, the amount of copper that had to be obtained from one ton of ore in order to make a profit changed from thirty pounds to ten, and today it may be as little as seven. Numerous mines these days are open pit operations that may yield 100,000 tons or even more per day of intermediate or low-grade ores. Only a few years ago such large but low-grade deposits would have been economically unworkable.

Particularly in the case of copper, there has been a tendency in recent years to concentrate on working open pit mines that are generally cheap to operate, and to stay away from more costly underground operations. But as the open pits have become deeper and the grades of ore near the surface have become lower, interest is again being shown in underground mines. Recent discoveries of copper deposits and statements by mining executives make it clear that underground mining of copper ores will accelerate in the future.

Many other mineral commodities are obtained from underground mines, which are gradually becoming deeper and deeper. Several properties, especially gold mines, are now working at depths of more than 10,000 feet.

Oil and gas explorations are also proceeding at ever greater depths; many holes have been drilled to more than 20,000 feet, though few of these very deep holes have been productive. In July of 1971, Texaco, Inc., announced the completion of the world's deepest producing well to date. From this well, in west Texas, flowed twenty million cubic feet of natural gas a day from a depth of between 21,837 and 23,040 feet. The field, the Gomez or Ellenburger, had in July of 1971 about seventy wells producing from depths greater than 20,000 feet, at an average development cost per well, it is said, of about one and a half million dollars. The deeper oil fields commonly produce more gas and relatively less oil than do the shallower fields.

Indestructibility of metals

The life of a durable metal is indefinite, almost certainly longer than that of any other commodity. For example, the annual production of gold is said to be only about 4% of existing stocks. Gold that was in use during the early history of mankind is probably still being used. Some of our present-day gold could have reposed at one time in the coffers of ancient Egypt or among the treasures of the Incan Empire; certainly it was never destroyed. During the course of time some gold is lost; every once in a while someone turns up a buried treasure or recovers a sunken cargo of Spanish doubloons from the depths of the ocean where they have lain for four hundred years or more. Then, too, minute quantities of gold are worn away, for gold is soft. Wedding rings become a little thinner as time passes, just as coins become slightly worn through constant handling and small amounts of silver and copper are removed when tarnished metal is cleaned. Iron rusts and slowly deteriorates, but even rusty scrap iron may be reused. Metals do not disappear under normal circumstances. They may be reclaimed and used again and again.

Fig. 2-3. Scrap metal recovered in the United States in 1971. Source: United States Bureau of Mines.

Metal	*Short tons*	*Percent of U.S. consumption*
Iron	39,300,000	28%
Lead	500,000	35%
Copper	467,000	24%
Aluminum	180,000	4%
Zinc	70,000	5%
Chromium	70,000	18%
Nickel	42,000	21%
Tin	14,500	19%
Antimony	18,600	61%
Magnesium	3,000	3%
Mercury	419	21%
Tungsten	250	3%
Tantalum	96	16%
Cobalt	34	1%
Silver	939	21%
Gold	91	30%
Platinum metals	11	25%

Since metals do not easily wear out, they accumulate. Numerous junkyards around the fringes of towns and cities are marked by unsightly heaps of old automobile skeletons and piles of dirty scrap metals. The fact that these automobile graveyards constitute a backlog of considerable value in times of emergency hardly compensates for their appearance.

During the 1920's the prediction was made that the world might actually reach a position where its needs for several metals might be met from stocks in use, supplemented by only minor additions of new metals. Obviously this is impossible under present conditions of improving standards of living and rapidly increasing demands. Such a prediction presupposed that demands would remain static and that no new uses for the metals would be found, which is being disproved almost daily.

Also, there is a great difference in the quality of scrap metal. *High-grade scrap* contains a high concentration of a metal. In the case of iron, it is heavy, as in railroad rails or automobile engines. It has a small percentage of impurities. *Low-grade scrap* contains only light-weight sheets of a metal, as in automobile bodies, and may have impurities that are difficult to remove and make the metal less fit for certain uses. For example, even minor amounts of impurities cause copper to become unsuitable for electric wiring because they decrease its ability to conduct electric currents.

The presence of large amounts of high-grade scrap metal available for use exerts a profound influence on the metals markets. When the price of metals goes up, the price of scrap generally follows, and increased and profitable efforts are then made to collect and sell it. Thus, at higher prices more scrap will reach the market than at lower rates, and an abundant supply will help to keep the price of the primary metal steady. A well organized scrap metals market is highly desirable because it stabilizes the primary metals prices.

Many people who thoroughly dislike the unattractive scrap yards and want to clean up around towns and cities do not understand why we need to mine large amounts of new metals when scrap is readily available. Why cannot it all be recycled and reused? The answer is strictly economic; this is not profitable, especially with old automobiles. The sheet metal of automobile bodies is thin, with little bulk, and before it can be used it must be pounded into compact pieces. It must be collected and hauled to a central yard, which takes energy and is expensive. Likewise, many alloys, set aside as impurities in the steel mills, must be separated almost entirely by hand, and the cost of this work is prohibitive. In spite of these difficulties, recycling is increasing. More scrap is

Fig. 2-4. Consumption, in short tons (except silver in ounces and excluding coins, and tin in long tons), of certain metals in the United States. Source: United States Bureau of Mines.

| | Primary metal | | Scrap | |
	1960	*1970*	*1960*	*1970*
Pig iron . . .	66,626,336	90,126,000	66,468,708	85,559,000
Copper . . .	1,349,896	2,070,000	1,208,434	869,400
Lead	1,021,172	1,350,000	469,903	597,390
Zinc	1,158,938	1,198,000	265,820	339,527
Silver	151,007,000	250,000,000	49,007,000	60–80,000,000
Tin	51,530	58,027	29,030	20,802

being used than people generally realize. For example, more than 40% of the copper now being fabricated in the United States is reclaimed metal. We should encourage the recycling of all types of scrap metals wherever this is economic. The recent establishment of collecting centers, especially for old cans and small bits of metal, should be widely expanded.

Dealing with large amounts of scrap metals, however, requires the installation of heavy equipment, because it cannot be done by hand. This equipment necessitates the use of more metals and considerable energy, and contributes to environmental pollution.

Substitutes

When commodities become too high in price or too scarce for ordinary uses, substitutes are sought and often found, though they are not always as satisfactory as the original material. For instance, as the price of copper goes up and the metal becomes in short supply, aluminum and plastics are substituted wherever possible; aluminum has proved to be an excellent substitute for copper in high tension transmission lines. As silver grows scarce, nickel and copper take over in "sandwich" coins and stainless steel is used in tableware. As tin becomes scarce, plastics or other materials become adequate substitutes. Generally, however, the use of a substitute means the loss of some desirable quality, an inconvenience, or an increase in cost.

It is of interest to note that the long-term production records do not show a single instance in which substitutes have totally displaced a material, leaving it without use. Each mineral has some use for which it is pre-eminent, and new uses are continually being found while old ones are extended. The substitution of aluminum for copper in high tension transmission lines, while releasing large amounts of copper, has not resulted in the closing of any copper mines; the use of copper has greatly expanded and the demand continues to grow at a rapid rate. As a matter of fact, if copper were to be used now instead of aluminum in transmission lines, severe shortages of copper would develop in other fields.

As we have already pointed out, adequate substitutes are not always available. Mercury, the only metal which is liquid at ordinary temperatures, has special uses for which nothing else has been found.

While more and more substitutes will be sought in the future and will help to reduce impending shortages, it should be borne in mind that many of the substitutes contain or are made of other nonrenewable raw materials. Sometimes people emphasize the use of plastics without realizing that plastics consume large amounts of petroleum as their base as well as other scarce materials such as fluorine. Plastics also require the use of heavy machinery for their manufacture, machinery that requires metals and energy, and the energy and waste products of processing may add to the pollution of the environment.

Ocean mining

As the need for minerals grows and as some become difficult to obtain, we look more and more to the seas as a source of supply. Many extractive companies are studying the possibilities of the sea—sea water, the sea floors, and mining in the bedrock beneath the oceans. Much of our petroleum and considerable amounts of other mineral resources are now being recovered from the shallow margins along the borders between the deep ocean basins and the continents. Very little has yet been done in the deeper waters.

Who owns these waters? Who will regulate the mining, and how will property and mineral rights be established? How far do territorial rights extend from the land into the oceans? These difficult and perplexing questions are now being studied by the legal profession in a number of countries and under the auspices of the United Nations, but it may be years before they are satisfactorily resolved to the point where the necessarily large capital investment will be encouraged to go into ocean mining. Perhaps developments in the relatively shallow waters of the North Sea in the search for oil and gas may establish precedents and resolve some of these questions.

While we may some day find in the oceans many of the minerals we need, their recovery is economically feasible at present only in the shallow coastal waters. Manganese is a good example of a commodity that cannot now be recovered. Nodules of manganese oxides that also contain appreciable amounts of nickel and copper are widely distributed in the deeper ocean basins. Numerous proposals have been advanced to recover these nodules, but so far they have been unsuccessful; the nodules are too low in grade and their mining and metallurgy are still too difficult and expensive. But to the extractive industries the possibility of their recovery remains a challenge, and ocean mining offers possibilities for the future when technological problems are resolved.

Prices and costs

The prices for most of the commodities of the extractive industries vary almost from day to day, depending on supply and demand, on the quantity purchased, on the place of sale, and on other factors. Therefore it is necessary, in considering these prices, to look at them over a period of time as long as a decade. The United States Bureau of Mines' *Minerals Yearbook* lists figures for 1960 to 1970, as shown in Fig. 2-5; anyone who wants a recent price for comparison may get it from nearly any metropolitan daily newspaper or from the *Wall Street Journal* or the commodity's trade publication.

In looking at the figures of the table, notice the fluctuating and cyclical character of metal prices. Most of these prices are f.o.b. (that is, freight extra) the center of production, such as St. Louis for lead and Ontario for nickel.

Like their prices, the costs of the extractive industries differ from those of other industries. Because most of the extractive industries are not primarily retailers, their service cost is less than that of the average business. They do not pay as much for advertising, service, and style. The price of a mineral commodity is therefore basic, since it includes little or none of the fringe costs that are attached to the price of a retail product—the cost of middlemen, salesmen, and advertising, the cost of dressing a product in an attractive package and projecting its image upon the public consciousness, or converting it from what the farmer has grown to what the city consumer will eat. When we consider the economics of the extractive industries, we are considering principally two costs:

• Cost of exploration, development, extraction, transportation, and smelting.
• Cost of the capital risked.

Thus when we talk of the costs of an extractive industry in dollars, we are talking of the cost in actual energy expended plus the cost of the prior expenditure of energy that produced the risk capital. The industry's cost is basic, its price is low, and its margin of profit is small.

As costs of exploration and development rise, as demands increase for minerals that become harder to find, wise governments tend to recognize the risks involved, to encourage the search for new deposits and the development of new tools for exploration and extraction, and to permit people willing to risk their capital a fair opportunity to recover their investment plus a profit—a greater profit than could be earned in a savings account, so that it warrants the hazard of loss. This would seem justified in consideration of the need served and the contribution made to national welfare by the extractive industries.

Other economic factors

In addition to the economic factors peculiar to the minerals industries as a whole, there are economic factors peculiar to each mineral commodity that do not apply to other minerals or their products. Because of the individual nature

Fig. 2-5. Comparison of metal prices, 1960 and 1970, in the United States. Source: United States Bureau of Mines.

Commodity	1960 high	1960 low	1970 high	1970 low
Copper	$ .33	$.30	$ .60	$ .53
Lead	.12	.11	.165	.135
Zinc	.13	.12	.155	.15
Tin	1.0475	.99675	1.88	1.605

of raw materials, the extractive industries in the past specialized in a single field—one was a copper company, another an energy company or a lead company, for instance, but none dealt in a combination of commodities. There are byproducts in any extractive industry. Silver and gold, for example, may be found in a copper deposit, and they have been mined and sold because of their economic value, although their quantity was too small to permit specialization and the company mining the deposit remained fundamentally a copper company.

The modern trend, however, is toward diversification. Many extractive industries now deal in a variety of products. Kennecott Copper, for instance, produces lead, zinc, silver, gold, bismuth, lithium, and coal as well as copper. St. Joe Minerals, which used to be St. Joe Lead Company, has added bismuth, copper, silver, gold, petroleum, gas, coal, and iron to its lead and zinc business. In diversifying, these companies must deal with the separate problems presented by the different minerals.

Uranium is an excellent example to show that each metal has its distinctive characteristics and problems. In the treatment and recovery of uranium, precautions against radiation are necessary, and the disposal of waste liquors from nuclear reactors is difficult and dangerous. In spite of their fine safety records, the location of nuclear reactors is a matter of great public concern. Even in the underground mining of uranium ore, small amounts of radon gas are generated, and in the United States the Public Health Service has established rigid restrictions on the amount of gas permitted in the air in places where men are working.

As uranium is strikingly different in some respects from other mineral commodities, so each raw material has properties that distinguish it from others. The special characteristics of oil and gas are very unlike those of other materials. Clays differ among themselves, and their physical properties are in general different from those of the metals. When gold is considered, we must realize that a small amount is worth a great deal of money, that until recent years the principal market has been government, that the price has been fixed by law, that the metal is practically indestructible, and that people everywhere hoard it. The characteristics of gold could hardly be more different from those of iron, which is among the cheapest of metals.

Each mineral commodity has its own special features and peculiarities that influence milling and concentration, use, and methods of mining. In the next five chapters we consider in detail some of the most important mineral commodities. In the case of each commodity, we describe its importance and use, where it is found and in what abundance, how it is obtained or produced, its economic aspects, its political aspects, and the outlook for its future.

A partial list of household uses of some nonrenewable natural resources

Alumina –   Abrasives: "sand" paper, knife-sharpening stones, natural or artificial rubies and sapphires worn as gems. "Jewels" in watch works.

Aluminum –   Window and door frames, screens, food containers and wrappers. Vacuum cleaners, floor polishers. Pots and pans and a host of kitchen gadgets. Roof covering, house sheathing, nails, waterproofing.

Arsenic –   Insecticides: as lead arsenate, calcium arsenate and Paris Green (arsenic trioxide plus copper acetate).

Asbestos –   Used as a thermal and electrical insulation. Sheets of asbestos are wrapped around the heating and cooling ducts of the home. Is used in the cords of hot appliances such as electric irons, heaters, coffee pots. Used as a filler in asphalt and vinyl plastic floor tile. Used in roofing tiles and exterior house finishes. Automobile brake blocks and clutch plate facings.

Barite –   Paint pigment and a paper filler to make smooth heavy paper for half tone color printing.

Beryllium –   Alloyed in small amounts with copper which it hardens and strengthens. Used as springs that actuate the breaker points in the distributor of your automobile. Used as a coating inside fluorescent light tubes.

Chromium –   Non-corrosive plating mostly on steel for ornamental or aesthetic reasons (automobile trim). Stainless steel alloys for tableware, pots, pans, kitchen sinks, cutlery, and other kitchen gadgets. Rust-preventive paint to protect steel on the outside of houses (as zinc chromate).

Clay –   With limestone to make Portland cement. Ceramics of a wide variety: fine china like Limoges or Rosenthal; common porcelain bathroom fixtures; bathroom and kitchen tile; pottery, stoneware, sewer pipe; common and fancy bricks; ceramic glazes;

Coal –   Space heating. Not much used in the west. Use is declining elsewhere because of air contamination by smoke and sulfur dioxide. Gasification of coal is coming into use as a substitute for natural gas. Coal fired power plants furnish much of the electricity used in the home.

Cobalt –   In the permanent magnets of loudspeakers of radios, TVs, HiFis. In alloy steels. Deep blue glass and pottery glazes.

Copper –   Wire, in all the electric wiring of the home for light, heating, power, telephone, radio, television, the motors in electric powered appliances. Copper pots, or at least copper bottoms on pots. Household ornaments like ashtrays. Brass hardware is part copper. Bronze hardware is part copper. Used in door locks, hinges, knobs. Bronze faucets (chromium plated) in the kitchen and bathroom. Bronze or brass machinery in the toilet tank. The average American automobile contains 20 pounds of copper as wire for the electrical system and for the radiator. "Better" homes have copper water pipes.

Diamonds –   Are a girl's best friend, and make the best HiFi needles.

Feldspar –   Porcelain, porcelain glazes, bathroom fixtures. Scouring powder. Non-returnable bottles.

Fluorine –   Essential constituent of teflon on cooking wares. Ingredient of Freon, the gas-liquid used in refrigerators, air conditioners and aerosol bombs. Used in some toothpastes to prevent cavities. Additive in drinking waters to prevent tooth decay.

Garnet –   An abrasive—the red colored "sand" paper for refinishing furniture.

Gold –   In dentistry. For aesthetic purposes: Jewelry, watches and other ornamental uses. In electronics where a non-corrodible conductor is needed.

Graphite –   A dry lubricant for door locks, or used as an additive to automobile greases and oils. It is used in blocks as the "brushes" of small universal motors such as are in vacuum cleaners. It is the central black bar or electrode, of the common flashlight batteries. It is the "lead" in a lead pencil.

Gypsum –   It is the principal constituent of the interior of modern houses finished inside with wall board or plaster. Used as a soil conditioner in the garden.

Iron –   Cast iron bathtubs, gas furnace burners, frying pans.

Steel –   Fasteners: nails, screws, nuts, bolts. Tools: hammers, saws, pliers, wrenches. Kitchen and tableware: pots, pans, knives, flat tableware (cast, pressed, stamped, plated, enameled, made of stainless steel alloys). Appliances: stoves, refrigerators, washers, dryers, garbage disposals, air conditioners, furnaces, air ducts, many others. Automobiles: One or two tons of steel. Decorative: wrought iron railings, curtain rods, furniture.

Limestone –   Medicinal use. Manufacture of Portland cement for concrete. Roofing material. Cut stone—ornamental and monumental.

Lead –   Was once extensively used in households as pipe, solder, and in paint. Now it is almost eliminated because of its poisonous properties. Household use is about limited to solder in the electrical system where it presents no hazard. Used in tetra-ethyl lead additive to gasoline for its anti-knock properties. Used in caulking the drain plumbing of the house. Most important use is in the lead-acid storage battery of the automobile.

Magnesium –   Used alloyed with aluminum. It may be in your window and door frames. It is surely in the aluminum step ladder, and if you own a Volkswagen it is a principal constituent of the aluminum engine. Medical purposes: Phillips milk of magnesia (magnesium oxide) and Epsom salts (magnesium sulfate). Metallic magnesium in fire crackers and flash bulbs.

Manganese –   In special alloy steels used in automobiles and household appliances. Manganese dioxide is the black stuff in ordinary flashlight batteries that you also use in small radios or televisions and children's toys.

Mercury –   Thermometers. Silent electric light switches. Mercury-silver batteries for small radios, cameras and hearing aids.

Mica –   Insulation in electrical apparatus such as electric irons, vacuum tubes and other electronic apparatus.

Molybdenum –   In alloy steels in the family automobile and in appliances. In kitchen cutlery.

Nickel –   Coins. Protective plating, now largely displaced by chromium which is whiter and brighter. Constituent of stainless steel. Nichrome wire heating elements in electric heaters. Permanent magnets (with cobalt) in the loudspeakers of radio, TVs, and HiFis.

Petroleum,
gas, & oil –   Gas used for space heating, water heating and cooking. Gas is the basic raw material for most plastics. Oil must be refined to be useful. It yields many products, many of which are used about the home: gasoline, kerosene, automobile engine oil, automobile greases, vaseline, parrafin, medicinal oil, asphalt for floor coverings, roofing, pavement, building papers.

Phosphates –   Fertilizer. Water softener. Detergents.

Platinum –   No necessary household use at present. May appear soon in catalytic after-burners on the family automobile to reduce smog. Used in jewelry. A matter of ostentation since it is not as handsome as silver, but is much more expensive.

Potassium
(potash) –   Fertilizer. Soft soaps.

Sand &
gravel –   Concrete when mixed with Portland cement: foundations, floors, and outdoor walkways of homes; driveways and street pavements. Asphaltic concrete when mixed with hot asphalt or emulsified asphalt to make pavement.

Selenium –   The active photo-electric materials in some camera exposure meters, and the electric eyes that open the store doors for you. Also serves as a paint pigment in orange, yellow, red. Used to color ruby red glass. Important in dry plate rectifiers in radios, TVs, etc.

Silica
(the oxide
which is
quartz) –   Basic material for glass for windows, jars, bottles. Used in ceramic glazes. Used as a gemstone—amethyst, citrine, Brazilian topaz. Used in sun lamps to give you an indoor tan or to cure a dermatological condition.

Silicon
(the metal) –   Used in transistors for miniature radios, recorders, tape players, etc.

Silver –   Photographic film and prints. Electrical contacts in timers and switches of automatic appliances. Aesthetic use as tableware and hollow ware, and jewelry, plated or pure (sterling). Coins. Dental and medical use. In mercury-silver batteries (see Mercury, above).

Sodium
(compounds) – Salt-cooking, water softening, melting ice, making ice cream. Sodium carbonate—used in cleaners. Sodium bicarbonate—for cooking. Sodium hydroxide—for making soap. Soda ash—important constituent of glass.

Strontium –   Makes "red fire". Used in the highway warning flares that all of us should carry in our cars for emergency use.

Sulfur –   A fundamental chemical to make sulfuric acid used to make an enormous variety of things. Sulfuric acid is the liquid in the storage battery of your car. Pure sulfur powder is used in the garden as an insecticide and to prevent mildew. It is a constituent of gypsum.

Talc –   Cosmetics, ceramics (dinner ware). Rubber tires, dry lubricant.

Tin –   A thin plating on the "tin" can which is really a steel can. Heavy plating on some kitchen ware such as meat grinders. Used extensively in restaurant kitchen ware. Alloyed with copper to make bronze for hardware and plumbing. Alloyed with lead in solder, used in: Home electrical systems. Sheet metal work about the house; roofing and drain plumbing. Automobile radiators and air conditioning and refrigerator machinery.

Titanium –   Paint pigment. Has the highest covering capacity and is the whitest white (not poisonous like white lead).

Tungsten –   Incandescent electric light filaments. Breaker points of automobile distributor. Contact points of thermostats such as control toasters, coffee pots, air-conditioners, furnaces, oven timers. In phosphors of fluorescent lights and TV tubes.

Vanadium –   In alloy steels in the family car. Increases strength, toughness, and hardness. Probably also in some household appliances.

Zinc –   Protective coating on steel—called "galvanized iron", used as sheets for gutters and downspouts, iron water pipes, nails for external use. Die casting alloy with aluminum in small machines such as automobile carburetors and fuel pumps; water pumps of automobiles and washing machines; speedometers and kitchen appliance timers, pulleys in washers and dryers. Zinc metal is the outer element of the dry cell-flashlight battery (it is covered with steel to prevent leakage). Brass-zinc alloyed with copper. Paint pigment—Zinc sulfide and oxide (not poisonous like lead). Used in manufacture of automobile tires.

3

Iron and Steel

Perhaps it was during the Stone Age that some early man, who may have seen a meteor fall, picked up a meteoritic fragment, pounded it with a rock, and found that he could hammer it into shape; here was a substance that would not fracture as rocks do. He may have made a spear point, a knife, or an axe from this piece of iron. He may have fashioned an ornament; we know that beads of meteoritic iron appeared in Egypt as early as 4000 B.C. They were worn by royalty and considered more valuable than gold. Meteoritic iron was also used for religious objects because early man could see that meteors came from heaven.

Probably iron was first smelted by accident. One of our primitive ancestors may have dropped meteoritic iron in a fire, or made a fireplace out of iron ore, which then melted. When he used this partially refined iron he could produce hunting tools that were stronger than any previously known.

However and whenever it occurred, once man discovered iron and learned to process the ore by fire, he began to use it more and more for all kinds of utensils. By 1000 B.C. he had learned to improve the substance by using a crude charcoal furnace to remove impurities (silicon and oxygen). As a result, he obtained "sponge" iron, which he could pound and shape. 1000 B.C. is usually taken as a convenient date to mark the end of the Bronze Age in Europe, Asia, and Egypt. Iron became the tool of conquest and the basis of wealth. The Iron Age began.

Through the intervening centuries iron has maintained its importance. Around 1700 the development of the blast furnace provided a continuous process to make metallic iron. Although the methods of smelting were not changed, they were refined to produce bigger batches of metal made in bigger containers. The cost of iron, which had been expensive, was reduced, and large quantities of cheap iron made possible the large quantities of steel that have formed the basis of industrialization. Steel became the most important of the three forms of commercial iron, far surpassing cast iron and wrought iron in usefulness.

At the present time, a country's heavy industry is the index of its material standard of living—and iron and steel are synonymous with heavy industry. The ambition of many a newly emerging nation is to develop its own iron and steel industries. Steel has become the symbol of modern civilization and the representation of power in international affairs.

Iron is one of the elementary substances of nature. It occurs in the native state only in meteorites, but its ores are abundant, the commonest being hematite (Fe_2O_3) and magnetite (Fe_3O_4). Both of these are oxides of iron, or compounds of iron and oxygen—one might say that they are natural bodies of iron rust. Iron is obtained by separating it from the oxygen.

In the strict sense, the element iron is scarcely an industrial material. Iron in that sense is, however, the principal constituent of cast iron, wrought iron, steel, and many other metals that contain small to large amounts of other elements. These other elements, called collectively alloying elements, give the metals highly distinctive values, properties, and uses. Thus "iron" as it comes from the blast furnace is a crude metal containing a varying amount of carbon. After it is cooled to the solid state it is called pig iron. Pig iron, which is iron that has been smelted, can be made into cast iron, wrought iron, or steel by controlling the amount of carbon, by tempering, and by other methods of working the metal.

Cast iron, made by pouring the metal into a mold after it has been smelted, is iron containing a relatively large percentage (between 2% and 6%) of carbon; it is used in making machine parts, stoves, radiators, and pipes. White cast iron, which is hard and brittle, results when the mold cools rapidly. If the mold cools slowly, part of the carbon separates itself as graphite, and the remaining metal is gray cast iron, softer and less brittle than the white. Cast iron is an important industrial material.

Wrought iron, made by reworking the metal after it has been smelted, is almost entirely free of carbon (0.1% to 0.2%). It is malleable, corrosion-resistant, ductile, and soft, and is used for bolts, rivets, water pipes, anchors, and chains. Being easy to rework, it is popular for ornamental ironwork and was the commonest form in which iron was used during the Middle Ages.

Steel is iron plus the alloying elements in amounts controlled so as to yield metals with predetermined properties that depend not only upon the amount but upon the kind and combinations of alloying metals added. If the carbon content of the steel is not more than 1.5% the material is called plain carbon steel, which comprises the bulk of all steel produced. The addition of very small amounts of alloying elements to carbon steel can completely alter its properties. Phosphorus will make it brittle, molybdenum will make it tough, able to resist fracturing under repeated stress, and tungsten will enable it to hold a cutting edge at high temperatures. Other alloying elements added to steel give it other properties, such as hardness, stiffness, ductility, resilience, resistance to fatigue, resistance to corrosion, and resistance to abrasion. Several thousand different alloys of steel have been made.

No other common material is as strong as steel. No other substance is as essential for the construction of the large, heavy structures of modern civilization—skyscrapers, bridges, and ships. Aluminum bends easily. Concrete not reinforced with steel tends to fracture. Because of its strength, steel has more uses than has iron. Cast iron is brittle, as anyone who has dropped a cast iron frying pan may know from his own experience, and wrought iron lacks the rigidity as well as the tensile strength of carbon steel.

There are many uses of steel, covering thousands of items from buildings to writing pens. This wide usage is due in part to the great variety of alloy steels. The current development of rockets, jet engines, and new types of electronic equipment has given impetus to the development of new alloy steels especially designed to meet the specific requirements, such as a high melting point, of the new technologies.

Cast iron and steel, the food of modern industry, are both cheap, cheaper than the cheapest hamburger. At $300 a short ton, steel sells for about 15¢ a pound; it would be difficult indeed to find pet food that inexpensive. If iron and steel were as expensive as most metals, modern industrialization could never have developed to its present magnitude.

Where iron ore is found

It is fortunate for today's civilization that iron is the fourth most plentiful element in the earth's crust and that rich iron ores are widely distributed. We in the United States have been particularly fortunate in having readily available, at the surface, the very large and high-grade deposits of the Lake Superior region, including those of the Mesabi range. We have used them lavishly. They have provided for the building of our transportation and structural industries, and supplied the steel for two world wars. About three billion tons of ore have already come out of the Lake Superior area, and as a result the higher grades of ore are gone. Large deposits in other parts of the world that were operated prior to World War II included those in Europe—Sweden, Russia, Norway, Germany, France, Spain, and Great Britain—and others in Africa, India, South America, and Manchuria.

Following World War II, the needs for iron and steel throughout the world

Fig. 3-1. A flow sheet—from iron ore to iron and steel products. The interrelations of the requirements of various mineral industries are evident. Sources: Cordell Durrell and the United States Steel Corporation.

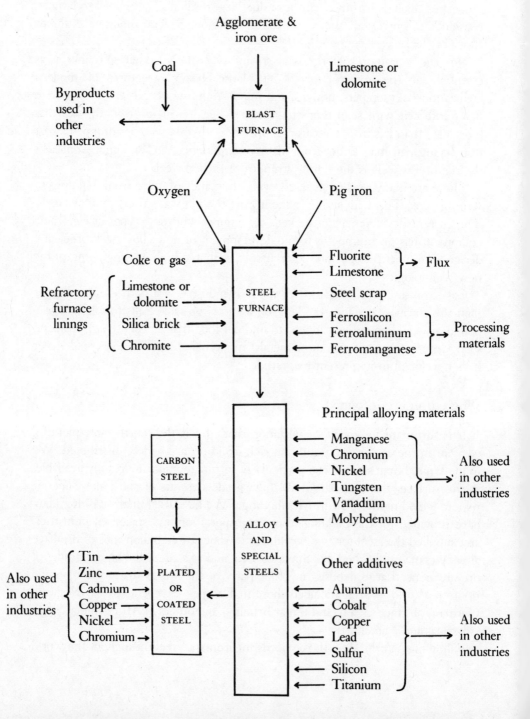

Fig. 3-2. The chief sources of iron ore concentrates and agglomerates in 1970. "Others" include Liberia, 3%; Venezuela, 3%; Great Britain, 2%; Chile, 1%; Mauritania, 1%; Peru, 1%; miscellaneous sources, 12%. Total production was 754,299,000 long tons. Source: United States Bureau of Mines.

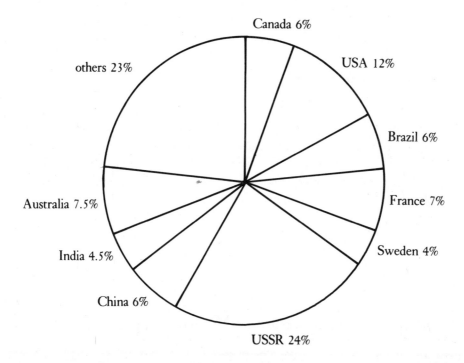

expanded at tremendous rates because of shortages caused by the war and also because of increases in the world's population and widespread improvement in standards of living. Hence the search for new deposits of iron ore was pushed vigorously at that time, and the results were successful, far beyond the most optimistic predictions. Although the world now uses more than seven hundred million tons of iron ore annually, so many discoveries have been made that iron ore surpluses exist in numerous countries. Sufficient ore appears to be available for many years and the search for new deposits has been curtailed.

The areas where iron ore deposits have been discovered since World War II are of special interest because their development has been responsible for changes in established lines of commerce and the building of communities and industry in remote places where settlements did not previously exist. The four regions that have newly developed iron ore deposits are:

• Northeastern Canada, including the Labrador-Ungava Bay area, Cartier, Wabush Lake, Carol Lake, and places farther north in the Arctic regions.

• The west coast of Africa from Mauritania south through Angola, including the deposits of Liberia, Guinea, and Gabon, which are as yet only partly developed.

• Australia, especially the west and northwest, including the Hamersley region in the Pilbara and Mount Goldsworthy. These deposits have been intensively developed in recent years.

• Parts of South America, including Cerro Bolívar and El Pao in Venezuela, Marcona in Peru, Romeral in Chile, and several deposits in Brazil.

From iron ore to plain carbon steel

From the ore lying in deposits in these and other parts of the earth, iron makes a long and tortuous journey and undergoes more than one metamorphosis before it emerges as the steel of, say, the Verrazano Bridge or a skyscraper rising in Tel Aviv or Mexico City. The course of its travel is determined by numerous economic considerations, from the primary question of "Is this profitable?" to the new technological advances that are speeding it at less cost from the mine to the steel mill. The winds of politics also affect the progress of the international shipment of iron ore, which may be set back by tariffs or pushed forward by some form of national self-interest.

The following steps outline the journey, tracing the processes by which steel is obtained from iron ore.

1. The iron ore is removed from the earth at a mine.
2. The iron ore (or a concentrated product known as pellets) is transported from the mine to a blast furnace.
3. In the blast furnace the ore is reduced to iron.
4. The iron is purified and turned into carbon steel in a steel furnace at a steel mill.

We proceed to describe in detail each of these processes.

The mining of iron ore

Iron ore in the earth is in bodies of many sizes and shapes. The value of an ore body depends upon three factors: its percentage of iron, its size, and its location. If the percentage of iron is high enough so that mining can be carried on at a profit, the source material is an ore; this profit test is always part of the definition. Today an ore body may contain as little as 25% iron if it is large enough and can be economically concentrated at the mine before transportation so that its content is about 60% iron.

Location is an important matter because of the cost of transportation. In Gabon, in west-central Africa, nearly six hundred miles from the coast, there are iron ore deposits that contain more than a billion tons of material that averages about 60% iron, and even larger tonnages of lower-grade mineralized rock. These deposits are not being worked because of the cost of mining and transportation. A railroad would have to be built, as would also a mining camp and port facilities. Cheaper ore is available elsewhere; only by mining extremely large tonnages could the unit cost be brought down to a competitive level.

Most iron ore is obtained from open pits, although some underground mines are being operated profitably. Mining companies are currently studying methods of mining huge tonnages underground, for example in the Lake Superior region.

The processing of iron ore

The ideal iron ore that can be sent from the mine directly to the blast furnace is "high grade," containing 60% or more iron. The fragments are evenly sized, and the fewer fine particles the better, since "fines" in a furnace will clog the open spaces between the larger ore particles and many will be lost by being blown up the furnace stacks. The ore must be strong enough to resist crushing either in transit or in the furnace, it must be porous enough so that gases penetrate it readily, and it must have surface areas as large as possible so that the greatest possible quantity of the ore is available for attack by the gases of the reduction process (*39*).

Prior to World War II the minimum iron content of ore used in the United States was between 50 and 52%; now it may be as low as 25 to 35% if the ore is magnetite, which can be concentrated far more easily than hematite. Improvements in processing to permit the use of lower grades of ore and to make more uniform the texture of the ore charged into the furnace have yielded tremendous economic advantages. They have decreased the amount of slag (the waste material), decreased fuel consumption, reduced transportation charges, reduced loss of fine-grained ore and dust through the blast furnace stacks, and generally led to increased furnace capacities and lessened costs.

Although this concept of the best ore has been held for many years, it is only since World War II, with the innovation of pellets and improved sinters, that serious efforts have been made to attain it.

Pellets and sinters. The process of pelletizing is one of forming, from small particles, larger aggregates of uniform size and properties. The iron ore is crushed and the iron oxide particles are concentrated, generally by the use of a magnet, while the waste material, called *tailing,* is discarded on dumps near the mine. The magnetic iron oxide particles are then cemented together by a material such as a special clay called *bentonite* and baked, after which they come out in small nodules or pellets about half an inch in diameter. Pellets are a semirefined iron ore.

The use of pellets is growing at a rapid rate and has revolutionized the making of iron. More than 60% of the iron ore used in the United States, or about seventy-five million tons annually, is now in pellet form. Because the pellets are of uniform size and permeability, they are more readily attacked by gases. They melt faster and at a more uniform rate in the gas of a blast furnace than does broken, untreated iron ore in its irregular shapes and sizes, and the waste of the dust-size particles of iron ore is largely avoided. Transportation cost, always an item to be considered, is lower because the pellet plant is at the mine and thus waste material is left behind and only pellets are transported to the blast furnace.

Not all mines have pellet plants. Run-of-mine (or *direct shipping*) ore is put through a screen at the blast furnace, the large pieces being taken out and sent directly to the furnace. The fine particles (fines) below a certain size are removed and agglomerated by a heating process. They are then called *sinters* and are used as part of the charge in the blast furnace. Sometimes other materials, such as limestone, dolomite, and fluorite, are added to sinters so that they become self-fluxing and melt easily when combined with the charge of the furnace. A *flux* is a material that makes another material melt at a lower temperature than it normally would; the addition of a fluxing material lowers the melting point of iron ore, thus saving the money that would be spent on additional heat.

Unlike pellets, sinter must be made at the blast furnace. If it were made at the mine, it would break up in subsequent transportation because it is so friable and brittle.

In 1950 only about 15% of the iron ore used in the United States was agglomerated, either sintered or pelletized; in 1960 the figure had increased to 52%, and in 1970 to 60% in pellets alone. The use of both sinters and pellets continues to increase. This is not surprising when one considers that the cost of a single blast furnace is in excess of sixty million dollars, that pelletizing can increase furnace capacity 30% or more, and that both pellets and sinters reduce the cost of operation.

Iron ore of the future

The use of improved agglomerates has also meant that many iron-bearing rocks formerly too low in grade to be of interest have become good ore and are eagerly sought. One of these deposits is in the Atlantic City district of Wyoming, where a pellet-producing industry has been established. Old

Fig. 3-3. Production and consumption of iron ore and agglomerates, in millions of long tons, in the United States, 1900–1970. Where the lines diverge, the upper line shows consumption, the lower line shows production, the space between shows the growth of imports during recent years. Source: United States Bureau of Mines.

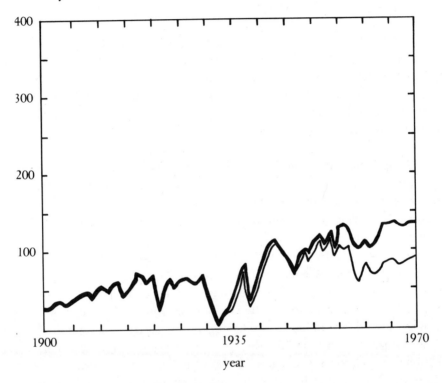

districts in the Lake Superior region have been rejuvenated as the low-grade ores, known as *magnetic taconites,* are now forming the basis of a rapidly expanding industry, the manufacturing of pellets. The demands of industry have changed radically since World War II; nonmagnetic medium-grade material formerly considered desirable is being disregarded, while low-grade rock that can easily be enriched is coming into demand. The increasing value of an ore body with less iron and the corresponding changes in values of ore bodies are typical of the changing conditions faced by the extractive industries. We have already pointed out copper as an example of the fact that as technologies are improved or invented to reduce the costs of mining and refining, the necessary mineral content of a deposit becomes less. This is true of almost all minerals.

It appears that the iron ores of the future will be of three types: 1. High-grade, direct shipping ores that contain more than 60% iron and have a minimum of fines. 2. High-grade but fine-grained materials that are suitable for forming into sinters, thus providing an excellent furnace feed. 3. Material of a grade and character that can be cheaply concentrated into fine-grained ore suitable for the manufacture of pellets.

The blast furnace

The modern blast furnace is a huge steel shell, about 130 feet high, installed in a vertical position and lined with heat-resistant brick. Its purpose is to reduce iron ore to metallic iron by separating iron from oxygen. It is to date the cheapest and most efficient method available. Once started, the furnace runs continuously until the lining needs renewal; it would be too costly in time and energy to let a furnace cool off, necessitating reheating before production could be resumed.

Precisely calculated amounts of ore, coke, and limestone are added at the top of the furnace and work their way down, becoming hotter as they sink. In the top half of the furnace, the coke begins to combine with the oxygen in the ore to form carbon dioxide (or monoxide), leaving metallic iron. Halfway down, the limestone begins to react with impurities in the ore and with ash in the coke to form slag, the waste material. The molten iron sinks to the bottom of the furnace, dissolving small amounts (up to 4%) of carbon en route.

This reaction is promoted by the concentration of air heated to between 1000° and 1700°F and enriched by combustible gases. After heating in huge stoves the air and gas are blown through openings called *tuyeres* into the bottom of the furnace. They roar up through the charge of iron ore, coke, and limestone, promoting the burning of the coke. It is the resulting gases that reduce the ore to metallic iron by removing oxygen from it, while the limestone causes earthy impurities to melt and flow, forming a fluid slag that floats on the molten iron in the lower part of the furnace. From the base of the furnace, 300 to 600 tons of iron are drawn off every three to five hours.

Molten iron and slag are the end products of the charge in the blast furnace. The molten iron drawn from the base of the furnace may be put into large containers which are taken directly to the steel mill without permitting the iron to cool, or it may be poured by a continuous process into molds, where it cools and forms bars, or "pigs." This pig iron is stacked and retained for future use as cast iron or for remelting at a later time to be converted into steel. Japan will buy all the pig iron it can get because "pigs" contain almost no oxygen and are therefore cheaper than iron ore to transport.

Slag has ordinarily been thrown away on a dump, but much of it is now being crushed and sold to the cement industry as a filler for concrete, where it takes the place of rock or sand in highway construction.

Further refinements in the process of the blast furnace include injections of natural gas, oil, or powdered coal to increase the temperatures and to speed the melting. Sometimes oxygen is added instead of air, and the top of the furnace may be pressurized to keep the gases in the charge for longer intervals. The blast furnace process may or may not have an effect on the environment (*15, 57*).

Efforts are constantly being made to omit the pig iron stage and replace the blast furnace by methods of forming steel directly from the ore, called *direct reduction* methods. Many of them have proved to be feasible; their costs cannot, however, compete with the costs of a blast furnace based on good coking coal. The blast furnace will continue to dominate the industry until such time as the costs of direct reduction (or some other process) are comparable.

Coking coal is a form of bituminous coal that is low in volatiles and high in fixed carbon. When this coal is distilled or partly burned in a semiclosed oven, it forms coke; the distilled gases are at times recovered and used in the chemical industry. Coke is a porous material that is strong and does not crumble easily. It is made up almost entirely of carbon and has many uses, one of the most essential of which is in the making of pig iron in the blast furnace.

Ideally, blast furnaces are situated near deposits of coking coal. A ton of coal occupies more space than a ton of iron ore; therefore, more tons of iron ore than of coal can be loaded into a railroad car or ship. Then, too, other industries are commonly based on coal fields and furnish a ready market for iron and steel products. Thus it is cheaper and more advantageous to ship iron ore to coal than coal to iron ore, and historically the largest and least expensive of the integrated steel-producing mills have been built near mines that produce coking coals. The Ruhr in western Europe and Pittsburgh, Pennsylvania, in the United States are outstanding examples.

In recent years, however, Japan has been able to build blast furnaces based on both imported iron ore and imported coking coal and to keep the costs so low that it can undersell all competitors. Because Japan must import the iron ore, coking coal, and most ingredients for steel making, and because transportation costs are so vital, Japan has concentrated its plants near its coasts to take advantage of the low cost of transportation by water. In 1971 Japan installed the world's largest blast furnace at the Fukuyama works of Nippon Kokan. This furnace has an interior volume of 148,000 cubic feet and a daily capacity of 10,000 metric tons of molten metal. Other nations, seeking the elusive goal of self-sufficiency and wanting to industrialize, have built steel mills near iron ore deposits far from the source of coking coal or other cheap energy, and where marketing potential is limited. Many of these furnaces are of questionable economic value and, if they are to be kept in operation, their output must be protected by tariffs. The outstanding success of the Japanese steel industry is influenced by the skill and energy of the people, and by the fact that they have taken full advantage of a large, expanding market and low transportation and labor costs.

Steel

The most modern steel furnaces use the basic oxygen process, which within the past few years has nearly revolutionized steel making as it has gradually displaced the older open hearth (reverbatory) process. In August, 1969, the output of the basic oxygen process in the United States for the first time exceeded production from open hearth furnaces.

The basic oxygen process utilizes a pear-shaped vessel, open at the top, with a capacity of about three hundred tons, into which molten iron and high-grade scrap are poured. A hollow needle is inserted into this bucket, going from the top nearly to the bottom, and through the needle a stream of almost pure oxygen is blown at supersonic speed into the molten iron. As oxygen bubbles up through the iron, it burns off carbon and makes of sulfur and other impurities a small amount of slag, which rises to the top and is easily removed. After removal of the slag, what remains is carbon steel.

An open hearth furnace is horizontal rather than vertical. The carefully calculated charge is added in the front of the furnace. Heat is reflected onto the charge, and the fuel does not come into direct contact with the molten iron.

In either the basic oxygen or open hearth process, the amount of carbon must be reduced, by the control of temperature, to less than 1.5%. A small amount of manganese must be added to the charge in order to make sound steel; no substitute is known that will serve for this purpose. Manganese acts as a *scavenger;* it combines with the impurities and most of it is then removed in the slag, taking the impurities with it. The basic oxygen process requires more of the mineral fluorite than does the open hearth process. Fluorite is used as a flux to make the slag more liquid, and demands for fluorite have increased as the basic oxygen process has become more common.

The basic oxygen process uses only small amounts of high-grade scrap rather than the low-grade scrap that can be used in the open hearth process to help start the melting. Therefore the increasing use of the basic oxygen process has decreased the need for low-grade scrap iron. This has contributed to the annual accumulation in the United States of some three to four million junked cars, which would cover some 30,000 acres of land, an area equal to that of the city of San Francisco (63).

The basic oxygen process takes only about fifty minutes or less to produce three hundred tons of steel. By comparison, the entire steel-making cycle in a conventional open hearth takes eight to ten hours, depending upon the original composition of the charge and the amount of steel produced.

Plain carbon steel, the product of the steel furnace (whether open hearth or basic oxygen), is used as structural steel or made into specialty steel in an electric furnace. In the large integrated steel mills, where steel is produced at lowest cost and the most diversified steel products are made, the molten and purified metal is not allowed to cool but is shunted rapidly from plant to plant until the final product is ready for shipment. In reality, these large integrated steel mills are intricate complexes of closely related industries. In addition to blast furnaces, basic oxygen or open hearth furnaces, and electric furnaces for the manufacture of specialty alloys, there are coke ovens, chemical plants, gas and electrical installations, and rolling mills and other massive equipment for the fabrication of shaped metals. The plants are designed to be as compact as possible in order to cut transportation costs and energy demands; the objective is to make the process continuous without the necessity of reheating cooled metal.

Most people who are unfamiliar with the operations of the steel industry have little appreciation of the tremendous cost of an integrated steel plant, which can easily run into hundreds of millions of dollars. Such investments are not made without intensive planning and thorough market studies. After they are constructed the mills should be operated at or near capacity in order to keep down unit costs and to remain competitive. Seldom is it feasible for small steel mills to manufacture more than a few shapes or items; their owners cannot hope to produce even a minor percentage of all the many steel articles in use today. Most of the manufactured goods must be left to the larger integrated mills or, more frequently, to the specialty manufacturer.

Economic aspects

Any country, if it is to be a part of industrialized civilization, needs plenty of iron and steel at low cost. These metals are essential for all large structures, for transportation, and for innumerable small items from cans to needles that are necessary every day in modern living.

To maintain low cost production of iron and steel there are several requirements.
• Not only regularly available supplies of good ore and coking coal, but enough risk capital to construct extensive, expensive installations and to provide for their continuous modernization.
• Operating costs, especially wages, fringe benefits, and taxes, that permit production at prices competitive nationally and internationally, and competitive with substitutes such as aluminum in cans or plastics for containers.
• Maximum use of productive facilities, which means that good markets must be available.

No modern nation can afford to do without iron and steel. Yet none can long afford to maintain a steel industry unless that industry is large and able to match competition, and is kept busy producing.

The steel industry in the United States has in the past had certain advantages.

• A large domestic market, as the twentieth century saw the development of the nation's automotive, construction, and transportation industries, all requiring steel.

• Fine supplies of good ore, such as the three billion tons taken from the Lake Superior district, and good local coking coals.

• Risk capital created by the country's growing industries and available for plant construction.

• Technology as competent as any in the world that, combined with domestic markets, made the United States steel industry competitive with that of any other nation.

• Capable management and productive labor.

Although the nation remains the world's foremost in per capita consumption, the steel industry of the United States today faces a somewhat different situation.

• We still have adequate supplies of ore and coking coal, but we now import 30% of the iron ore we consume, and in 1971 our imports of steel exceeded 18,000,000 tons (1).

• While our technology and plants remain good, the number of modern plants abroad is increasing, and some of them are superior to ours. Both the basic oxygen process and continuous steel casting were developed in Europe.

• Risk capital for new plants and modernization is not too easily attracted here because the return on it is relatively modest.

• Our iron and steel must constantly compete with substitutes for some uses.

• Our operating costs, including wages and fringe benefits, are the highest in the world. There are foreign steel industries, as in Japan, Sweden, and Germany, that have not only excellent modern plants but generally lower operating costs.

• We have largely lost our foreign markets to efficient foreign competitors. Once lost, a foreign market is lost forever in so competitive a field. In some cases, our loss of a market has been due to the fact that we persist in our system of measurement; when we join with the rest of the world in using the metric system, the resulting uniformity should prove advantageous to the United States steel industry.

• Production is at less than full capacity in our plants (in 1972, about 90%), which raises operating costs. Production is further curtailed when a strike occurs, with ensuing further rises in costs. Inflation of operating costs is a major factor in the industry.

• In 1971, for the first time in history, Russia produced more steel than the United States.

In spite of its present problems, the United States steel industry remains in a moderately healthy economic condition.

• This is partly due to the proximity of the industry to major domestic markets, which makes possible reasonable transportation costs and quick delivery. Detroit, for example, can buy the steel for its automobiles from Gary, Indiana, or from Pittsburgh, Pennsylvania.

• This is partly due to international politics. Japan and Germany might possibly be able to undersell United States steel makers in our country, and frequently do so on certain items. But all foreign nations know that if they do not restrain their exporters, our government might retaliate with import tariffs.

Over the world today, the picture of iron and steel is a bright one, brighter than in the United States.

• With continuing discoveries such as those in western Australia, ore supplies appear adequate for at least the projected needs of the twenty-first century.

• Reserves of coking coal, so necessary for steel making, seem sufficient for a long time.

• Technological advances continue.

Transportation costs are a factor of rising importance. It costs more per ton to send ore by rail to San Francisco from Fresno, a distance of about two hundred miles, than to ship ore to San Francisco from Perth, Australia, a distance of about eight thousand miles. For this reason, plants are more likely to be located near seaports, and the use of large freighters to transport ore is an economy now generally practiced by steel industries everywhere. Like the transportation of petroleum, this use is transforming the shipping business.

Diversification of ore supplies has been sought by the steel industry in all countries as a sound policy to lower the cost of raw materials and to establish a measure of protection against political upheavals, labor stoppages, monetary exchange difficulties, or unreasonable demands and unbusinesslike actions from any source, particularly where mines are in politically immature or unstable areas. The discoveries of iron ore after World War II gave the steel companies a fine opportunity to diversify their sources of supply, and as a result trade routes shifted from the former traditional sources as buyers became no longer dependent upon a single deposit or area. New supply lines have been established and stabilized, large investments have been made and are being amortized. Companies in the old, well established mining districts are cutting costs in every possible way in order to remain competitive in a business which is now evolving in new districts and new countries. Brazil for years refused to permit development of its iron ore industry or to export ore, with the avowed purpose of holding its reserves for the future at higher prices, but since about 1960 Brazil has been bringing its very fine ores into the world market and in 1971 exported iron ore worth approximately $300,000,000.

In a further effort to reduce the cost of raw materials most of the large steel-manufacturing companies have gone into the mining business, maintaining so-called "captive" mines where output can be geared to consumption.

The development of a major iron ore deposit usually costs many millions of dollars, and several, such as that at Mekambo in Gabon, may well cost more than three hundred million dollars. These vast sums of money have been provided from the resources of many companies, individuals, and governments, and have at times led to interesting economic situations. For example, American capital has developed a large iron ore industry in Australia to supply ore to Japan, where it will be made into steel to compete with American products. Private capital is developing ore deposits that compete directly with government-owned and -operated mines, or with programs supported by government subsidy.

Because of the very high cost of bringing a deposit into production, ownership is frequently divided among several companies and nations. An excellent example of the kind of organization now being formed is furnished by the group called LAMCO, which financed the development of iron ores in the Nimba range of Liberia. This venture is controlled 75% by LAMCO, the Liberian-American-Swedish Minerals Company, and 25% by Bethlehem Steel Corporation. The Liberian government owns 50% of LAMCO; the other 50% is owned by Liberian Iron Ore Company, Ltd., a Canadian corporation, which in turn is owned 70% by the Swedish Lamco Syndicate (a group of six Swedish companies, including the operating company in Liberia) and 30% by groups and individuals in Canada, Liberia, Sweden, and the United States (70).

Clearly the small miner and the independent middleman have no place in such a complex organization, and with few exceptions they are gradually being eliminated from large-scale international iron ore mining. This is a field dominated by groups with large amounts of risk capital and diverse technical skills, and by owners of smelters and fabricating companies who want to control their sources of supply. Such close control of ore output permits fabricators to tailor ore production closely to their day-by-day needs. Only Japan, among the large steel producers, depends upon ore purchases from independent mining companies.

LAMCO illustrates a significant modern trend: the willingness of large companies to join with governments and companies of other nationalities to form a truly international or multinational company. The result must be closer international cooperation, the political implications must be profound.

Political aspects

The importance of iron and steel in international affairs is self-evident. All countries want their own steel mills partly because they feel a need for self-sufficiency in order to maintain or improve their standard of living. This need became obvious during World War II, when many countries had their supply of iron and steel cut off for the second time in twenty-five years. In addition, iron and steel are a status symbol, and there are developing nations that have established the industry solely as a sop to national pride.

Political aspects of iron and steel are tied to economic aspects by such matters as taxation and tariffs. For example, Japan is able to produce steel and sell it on the western coast of the United States cheaper than can American steel producers, and as a result considerable agitation exists in the United States for tariff protection. During the wage and price freeze of 1971, a tax imposed on many imports from Japan was insufficient to make our own steels competitive; in the midst of the freeze, Bethlehem Steel Corporation announced its decision to abandon plans to establish a large integrated steel mill on the west coast.

Japan has its own merchant fleet of modern vessels, and transports its finished products, as well as much of its ore and coal, in its own ships. Although coal mines, steel mills, and shipping are all operated by private enterprise in Japan, the government takes an active part and its approval is required before any major action can be taken. The country is becoming increasingly involved financially with the conveniently situated and high-grade Australian deposits, but it still buys iron ore from a wide variety of places and avoids too great a dependency upon any one source of supply. Japan is now able to obtain coking coal in Australia or West Virginia and iron ore from South America, Australia, or even Nevada, to make steel which it transports and sells in California cheaper than steel can be made there, in spite of the vaunted technology of the United States. For political as well as economic reasons, Japan does not want to undercut the United States steel industry altogether. Japan would be hurt by protective tariffs against its exports, because the nation must export to survive.

Many governments subsidize the export of iron and steel in order that their mills may produce the volume needed if they are to be efficient.

A good example of how a progressive government can help a steel industry is shown by the implementation of the Schumann Plan for the steel and coal complex of the Ruhr in western Europe. This realistic plan was put into practice after World War II in an effort to revitalize the almost totally destroyed steel mills of the region. It brought together under one control the steel mills and the coal mines, and it further coordinated an already highly integrated complex, creating what is in effect a large, strong, and tightly held monopoly within the framework of the European Common Market. The Schumann Plan resulted in a greatly improved competitive position for the European steel industry, which is now one of the most modern and best organized steel complexes in the world.

Of all the industrialized nations, only Russia has sufficient iron ore for its long-range future. Russia also has excellent fields of coking coals and would probably expand production of its already large amounts of iron and steel if it could find a market. But Russia, because of the cost of transportation, will have difficulty competing with the highly efficient mills of Japan and western Europe.

The future

What is the outlook for the future of iron and steel? Per capita annual consumption of steel in the United States and western Europe is one ton; per capita consumption on a worldwide basis is less than one sixth of a ton per year. To raise the per capita consumption of steel on a worldwide basis to that of the United States and western Europe would mean that production would need to be increased to about 3.25 billion tons of steel annually. With these figures in mind, a study of needs projected into the future indicates that the upward trend will not be reversed. Substitutes will not replace the major uses of iron and steel; uses in the future will continue and even expand.

The Iron Age shows no sign of coming to an end.

4

The Ferro-Alloy Elements

The uses and value of steel are greatly increased by the addition to carbon steel of other elements, known as the ferro-alloy elements, that effect changes in its mechanical or physical properties. This addition of specific quantities of one or more elements produces what is known as an alloy, or an alloy steel. Some alloys that appear to be homogeneous reveal etched patterns when they are polished, and when viewed under the microscope they show intricate patterns made up of "microconstituents." Other alloys, known as solid solutions, appear to be homogeneous even under the microscope. A number of the better stainless steels are of this second type.

Most of the alloy steels are made in either electric furnaces or other steel-making furnaces where small amounts of material can be handled readily and where the composition and conditions of smelting can be closely controlled to meet exact specifications. The use of electric furnaces has grown at a very fast rate in recent years. Unlike the blast furnace, where only minor amounts of scrap may be employed as starters for the charge, the electric furnace is one place where a large amount of selected high-grade scrap iron is being used. As the reader will remember from Chapter 2, high-grade scrap is heavy and compact, such as railroad rails, while low-grade scrap is lightweight and thin and usually contains a considerable amount of alloyed metals, such as tin cans.

Not all of the ferro-alloy elements are metals. Carbon, boron, phosphorus, and silicon, also used as alloying elements, are nonmetallic.

For our purposes in this book, we use the following definition of "metal": A metal is an element or an alloy of elements. It has what is known as a metallic luster. It will ordinarily conduct heat and electricity better than will other elementary substances. As defined by a chemist, it carries a positive electrical charge.

The principal ferro-alloy metals are chromium, cobalt, manganese, molybdenum, nickel, niobium (columbium), tantalum, titanium, tungsten, vanadium, and zirconium. Some of the rare earth metals—cerium, lanthanum, and others—are also used in special steels.

The ferro-alloy elements serve for a great variety of products, but their primary use is as an addition to iron to form alloys. As the reader will remember from his study of iron, some of the properties given to iron or to carbon steel by the addition of other elements, often in surprisingly small amounts (less than 1%), are hardness, toughness, resistance of various kinds, and many other special characteristics needed by science and industry. The demands imposed by the rigid requirements of space exploration and research in atomic energy have given tremendous impetus to the development of alloys, and the science of materials in general has made startling progress in recent years. One of the real and lasting benefits of the "Age of Atomic Energy" is a better understanding and more extensive use of minerals, of which the ferro-alloy elements form an important part.

Some of the principal ferro-alloy elements have in general the following effects when alloyed with plain carbon steel. Chromium, nickel, and vanadium all make stainless steel. Vanadium, molybdenum, and manganese increase tensile strength and affect melting temperatures; molybdenum, moreover, adds resistance to corrosion. Tungsten makes steel hard and tough so that it can be used in tools to cut other metals. Cobalt also is used in the manufacture of high-grade tool steels. Niobium prevents intergranular corrosion in stainless steels. Since steels are made of interlocking grains, or particles, of metals, there are interstices where the metals meet which will corrode when the steels come into contact with acids. Niobium seals the interstices.

As we pointed out in the preceding chapter, the combinations of ferro-alloy elements in alloys with steels are numerous. Each large steel company publishes a listing of the properties of its alloys, of which there are several thousand. For instance, while both nickel steel and chrome steel are stainless steels, chrome-nickel steel may be a better stainless steel. Some steels may include four or five metals in various combinations originated by the companies that manufacture them.

The distribution of the ferro-alloy elements is erratic, nature having tossed them over the earth in an especially chaotic fashion. Except for titanium and nickel, all of the ferro-alloy metals are available only in limited amounts. Titanium, while abundant, does not often occur sufficiently concentrated so that it can be extracted at a profit. Nickel is plentiful in many countries in the tropical soils called *laterites* and in several sulfide deposits in Canada and Australia.

All industrialized nations import some ferro-alloy elements and for their supplies are commonly dependent upon the less developed nations. For example, the United States has a surplus of only one of the more widely used ferro-alloy metals, molybdenum, although our requirements of many of the others are partially fulfilled by domestic mining. The United States, therefore, must import a large percentage of what it needs, and so must Japan and the industrial centers of western Europe.

The need for ferro-alloy metals has caused considerable anxiety about sources of supply because of the questionable political stability of the countries of origin. It is the need for these metals that has forced the industrial centers to shape their foreign policies and trade agreements so as to encourage maximum stability in the developing nations, and foreign aid has also been directed with this need in mind. The ferro-alloy metals are of strategic and critical value both for industry and for defense. All of them, in one form or another, have at times been included in the reserves (or stockpiles) of the United States government.

No more statements about the ferro-alloy elements as a group are relevant at this point. It is time to consider them separately, since each has its individual uses, importance, occurrence, and economic and political significance.

Chromium

Chromium (or chrome), a heavy metal, is one of the best known and most widely used of the ferro-alloy elements. The United States at present manufactures about thirty varieties of stainless steel that contain chromium; stainless steel has a great and increasing use in the manufacture of cutlery and kitchenware. Chromium does not readily oxidize or tarnish—it does not form compounds with either oxygen or sulfur in the air—and therefore it retains its luster. In addition to its use in stainless steel, chromium has an esthetic and protective use as a coating for other metals, for example to decorate automobiles. Some 75% of chromium metal used in the United States goes into the manufacture of "super" alloys for the aircraft industry, while most of the balance goes into stainless and nickel-chromium steels. Chromium is also used in pigments (in chrome yellow, for instance), in dyes (where it imparts the color green), in textiles, in chemistry (where chromic acid is excellent for a pickling process), in the tanning of leathers, in storage batteries, and as a nearly neutral refractory. Considerable chromium is sold in the form of ferro-alloys such as high carbon ferrochromium, which contains 4 to 9% carbon and 65 to 70% chromium.

Chromium is obtained by the mining of chromite, its only ore mineral, which consists of iron and chromium combined with oxygen. Three classes of chromite are recognized depending on the approximate percentage of chromium in the ore.

- Metallurgical grade ore (47%), by far the most valuable, constitutes between 80% and 85% of all the chromium ore imported into the United States. It is the only kind that can readily be used for coating other metals and in stainless steel.
- Chemical grade ore (45%) is used in pigments and in the chemical industry.
- Refractory grade ore (34%) is used in furnace linings as a neutral refractory.

Fig. 4-1. Production of chromite in 1970. "Others" include Malagasy, 2.4%; Finland, 2.1%; Iran, 2%; miscellaneous sources, 3.8%. Total production was 6,526,894 short tons. Source: United States Bureau of Mines.

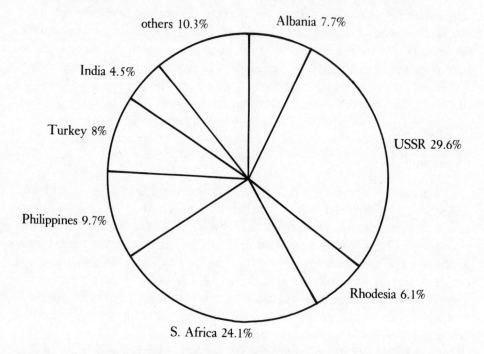

Chromium oxide is chemically neutral and does not easily react with other materials, even at high temperatures. It is therefore suitable for a furnace lining where a chemically inert substance is needed that will not combine with whatever is being treated in the furnace.

Metallurgical and chemical grade ores come principally from Rhodesia, Russia, Turkey, South Africa, and the Philippine Islands. Minor amounts are produced in Yugoslavia, Greece, Albania, and New Caledonia. Except for an insignificant quantity mined in California and a few other places, the Western Hemisphere has no commercial supply of metallurgical grade ore; all of it must be obtained from distant sources. Montana does have chromite near Stillwater; this ore, however, contains much iron and little chromium. The iron and chromium cannot be separated mechanically but must be melted or dissolved, which results in a process so expensive that the mining of Montana chromite has not to date been economically feasible. The only chromite deposits in the Western Hemisphere of continuous economic value for any long period of time are the refractory ores of Camaguey and Oriente Provinces in Cuba. Cuban chromite was sold principally to consumers in the United States until the properties, some owned by Cubans and some by foreigners, were confiscated by the Castro regime.

Fig. 4-2. United States chrome ore in long tons—the shifting pattern of sources of supply. Source: *Metals Week*, October 18, 1971, McGraw-Hill. Copyrighted and used with permission.

	1966	1967	1968	1969	1970
Rhodesia	73,000	50,000	—	—	—
USSR	155,000	165,000	183,000	164,000	225,000
South Africa	91,000	46,000	46,000	70,000	45,000
Turkey	16,000	21,000	29,000	35,000	64,000
Iran	—	—	—	6,000	15,000
Pakistan	—	—	—	—	15,000
Totals	335,000	282,000	258,000	275,000	364,000

The United States formerly imported much of its chromite from Rhodesia. When a United Nations embargo was imposed upon that country, forbidding the importation of its ore, industry in the United States was forced to buy both from Turkey and, in even larger quantities, from Russia. Russia, in order to profit from the situation, quickly trebled the price of its ore, which caused the United States Congress to reconsider the Rhodesian embargo. In 1970 one United States company was granted permission by the government to obtain 150,000 tons of ore for which it had paid prior to the embargo. By the time this transaction was approved, however, Rhodesia had sold the ore and the company was able to obtain no more than 10,000 tons (43).

In fact, the embargo did not directly hurt Rhodesia, which was able to sell its ore on both a short-term and a long-term basis. The purchasers and prices were not publicized but persistent rumors in the market place were that part of the ore was going to Russia for resale to the United States. In November of 1971 the United States Congress again authorized, by means of a rider attached to another bill, the purchase of chromite from Rhodesia, thereby breaking the embargo established by the United Nations. At that time it was estimated that not much ore would be available for three years, because the Rhodesian output had been sold so far in advance. Late in the autumn of 1973 the United States

Congress reimposed the embargo on all Rhodesian chrome ore. Is this action very different from the Arabian imposition of an embargo on the export of oil? Certainly both show the international influence of nonrenewable resources and the changing political problems of the extractive industries.

Cobalt

Cobalt is one of the most useful of the ferro-alloy metals, and no good substitute has been found for many of its uses. Until recent years nearly all cobalt was used in pigments (the lovely color of cobalt blue comes from a cobalt compound), to provide a glossy surface for ceramics, and as a dryer in paints. Later, however, additional uses were developed which have greatly increased the demand for the metal and thus its value. The most important modern use is in the manufacture of strong, permanent magnets, which are made of cobalt plus iron. Without cobalt, there would be no modern communications systems, for magnets are needed in radios, television, walkie-talkies, and many other electronic devices. A similar use is in the manufacture of magnetic tapes for recorders. Cobalt is used in medicine and also in the guided missile program and in the manufacture of jet aircraft engines, gas turbines, and turbo superchargers. Considerable amounts go into high-speed tool steels, hard facing materials, refractory alloys, and cemented carbides (carbides are compounds of carbon with a metal). Tungsten carbides and other cemented carbides are commonly sintered with 3% to 20% metallic cobalt binder.

Cobalt ores come principally from Zaire and Zambia, where they are byproducts in the mining of nickel and copper. In some other areas they are mined from nickeliferrous laterites. A small amount of cobalt was recovered in the United States as a byproduct of the iron ore mining at Cornwall, Pennsylvania. This iron ore has been worked out.

Substitutes such as vanadium, tantalum, and boron are at times used in place of cobalt; and nickel, though not so good for this purpose, can be substituted as a binder in cemented carbides. The fact remains that no substitutes have been found for the most important uses of cobalt and that cobalt is a relatively rare metal, much rarer than chromium, available in only a few places. The supply of cobalt is likely to remain limited unless the price is raised considerably, thus permitting treatment of materials that are now marginal.

Manganese

Manganese is a ferro-alloy metal essential to the manufacture of sound steel. Its primary use is as a scavenger, a deoxidizing and desulfurizing agent to extract oxygen, sulfur, and other impurities from steel. Between eight and fourteen pounds of manganese are used in the manufacture of each ton of steel, the manganese going off into the slag. No good substitute has as yet been found for manganese as a scavenger in steel making.

Manganese also alloys with steel to produce toughness and durability; it is used in the chemical industry, in potassium permanganate, and in drugs; and certain grades of it are needed for the manufacture of dry cell batteries. Only particular kinds of manganese will do for this last-named purpose and only empirical methods can prove whether or not any given manganese will serve—the manganese must be tried in a dry cell.

Manganese comes from manganese oxide ore; many minerals contain manganese, but only a few oxides and, in some places, carbonates are mined as ore. Three grades of ore are recognized: metallurgical for use in steel making and as an alloy, chemical, and battery.

The principal sources of supply are Australia, Africa (Zaire, Gabon, Ghana, and South Africa), Latin America (Brazil, Chile, Cuba, and Mexico), India, and Russia. Small amounts have been produced in the United States, particularly under the duress of war and with government subsidies, but at present all manganese used in the United States is imported. Russia probably has, at Nikopol and Chiatura, the largest reserves in the world, but very large reserves exist as well in Gabon and Australia. Prior to World War II, most of the manganese used in the heavily industrialized nations came from Russia, with lesser amounts from India and Ghana, and when, after the war, Russia refused to sell manganese to the western world, there was considerable anxiety in the steel industry over possible shortages of this essential material. Fortunately the shortages did not develop because, after an intensive worldwide search, large deposits were uncovered in Australia, Brazil, Gabon, and Mexico, and additional deposits found in Zaire and South Africa. As a result, the western world's dependency upon Russia for manganese no longer exists. India is said still to have considerable ore, but India has not retained its former share of the market, reportedly because of inability to meet contract obligations. The deposits of Ghana are said to be nearly exhausted.

Outside of Russia, which can supply its own needs, the principal users of manganese are the United States (in its industrial area), the countries of western Europe, and Japan. These nations, which are the large manufacturers of steel, are deficient in manganese and must depend upon imports. In 1970 the United States used 2.8 million tons of manganese ore.

The economic and political importance of manganese is illustrated by Germany's need for it during World War II to manufacture sound steel. When the Nazis invaded Russia, they drove their armies first of all, not toward Moscow, but toward the manganese center of Nikopol in the south. After they occupied Nikopol, they sent as much manganese as possible back to Germany;

Fig. 4-3. Production and consumption of manganese ore, in thousands of long tons, in the United States, 1900–1970. The lower line shows production. The upper line shows consumption. The divergence of the two lines shows the growth of imports. Source: United States Bureau of Mines.

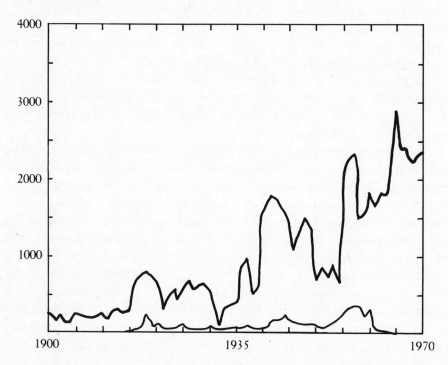

when they were forced to retreat, Nikopol was one of the last places to be abandoned. The shortages of manganese in Germany resulted in the manufacture of faulty, unsound steel that probably played a significant part in the defeat of the German armies.

All manganese ore prices are negotiated, depending upon the characteristics and quantity of ore purchased; high carbon ferromanganese contains 74–76% manganese. Specifications are set according to the use to be made of the manganese, which should generally run at least 46%, while silica and alumina together should run less than 6% or 7% and such elements as sodium, potassium, and arsenic should be low. Manganese ore is commonly quoted per unit, which is one percent of a long ton, or twenty-two pounds.

Molybdenum

Molybdenum is used principally as an alloying element in the manufacture of molybdenum steel, an excellent corrosion-resistant material which maintains high strength at high temperatures. Although comparatively rare in most parts of the world, molybdenum is the one ferro-alloy of which the United States has surplus quantities which this nation exports to much of the world. Considerable molybdenum is also recovered from the mining of copper in Peru and Chile and from several molybdenum mines in western Canada. Some is said to be produced at the Yokazyoshi molybdenum mine in Manchuria.

Molybdenum is mined alone, as in New Mexico, Colorado, and British Columbia, Canada; it is also a common byproduct of copper mining, as in Chile, Peru, the western United States, and western Canada, although it is not found in all copper mines. Three molybdenum mines in the United States, in addition to the production from copper mines here, furnish much of the ore for western industries. The largest molybdenum mine in the world is at Climax, Colorado; American Metal Climax Company has discovered uses for the metal as well as supplying it, and is a fine example of a company that mines a metal, develops its uses, and sells it. The same company is now developing another deep underground molybdenum mine at Henderson, Colorado, which may well replace the one at Climax as the largest and least expensive to operate. So much care has been taken in its development that conservation organizations have held it up as an example of what can be done to preserve the environment around a large mine. Another mine, somewhat smaller than those at Climax and Henderson, is in northern New Mexico between Questa and Red River.

Substitutes are available for most of the uses of molybdenum.

Nickel

Nickel, an attractive, glossy metal with a certain esthetic appeal, is a popular ferro-alloy element. It has many and varied uses that have resulted mainly from long-term, privately financed research and study. The largest amount of the metal is used in steel alloys to produce stainless steels, but a considerable quantity goes into high-temperature alloys, electric resistance alloys, and, especially in Canada and the United States, coins. There are also many minor uses, in nickel-cadmium batteries, for example. Nickel alloys with numerous other metals besides iron, as in nickel pewter.

The largest and by far the best deposits of nickel in the world are those of nickel sulfide in Canada, near Sudbury, Ontario, and at Thompson Lake, Manitoba. The International Nickel Company of Canada has long maintained a near monopoly of the metal and developed uses for it. Other places that produce considerable amounts of nickel are the island of New Caledonia in the South Pacific, Petsamo in the Kola Peninsula of northwestern Russia, and South Africa.

Early in 1969 the announcement that a large body of nickel sulfide ore had been found in western Australia precipitated a rush to the area. The ore body, discovered at a place called Kambalda, has proved to be an excellent one and is now in production. Several other bodies of nickel-bearing ores have since been found; the search continues and Australia is well on its way to becoming a large, steady producer of nickel.

Fig. 4-4. The growth of world nickel production, 1957–1970, in thousands of short tons. Source: United States Bureau of Mines.

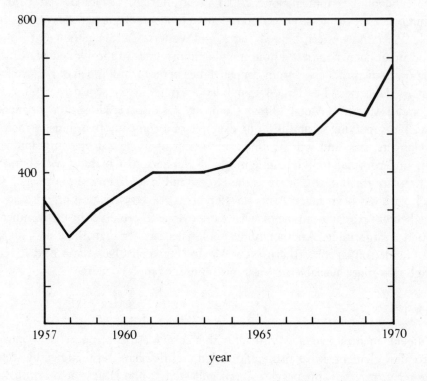

Comprising a tremendous potential source of nickel are the red soils commonly known as laterites that are widely distributed in many tropical countries. Nickeliferrous laterites form over rocks that contain small amounts of nickel, which is concentrated in the subsoil during weathering and in a few places can be recovered profitably. The metallurgy of the laterites is complicated, however, and the recovery of nickel from them is so costly that until very recently only a few of the deposits had been developed on a commercial scale. The minerals in the laterites are nickel silicates, more expensive to treat than the nickel sulfides of Canada.

Probably the best known and certainly the most productive lateritic ores are in New Caledonia, which furnished most of the nickel in the world before the discovery of the Canadian deposits. New Caledonia still produces considerable nickel, exporting most of it to France, but this source is of minor significance in the world market. Cuba also has large deposits of nickeliferous laterites, in part developed during and after World War II. Facilities for the concentration and extraction of the Cuban ores were being enlarged and modernized when the properties were expropriated by the Castro government. A small amount of nickeliferous concentrates is probably still being produced in Cuba and shipped to Russia or its European satellites for treatment and reduction.

Fig. 4-5. Nickel production in 1970. "Others" include the United States, 2.2%, and miscellaneous sources, 8.4%. Total production was 685,000 short tons. Source: United States Bureau of Mines.

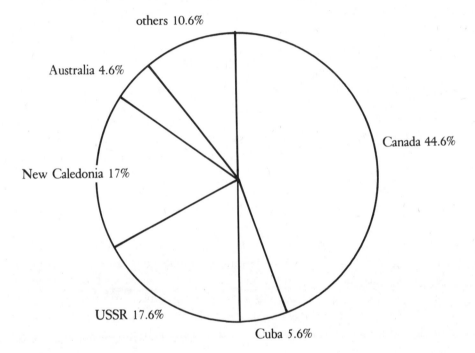

The laterites of Colombia and Guatemala have been partially explored, and development of nickel deposits in both countries is now going forward. Reports from the Philippine Islands indicate that the large deposits in the southern part of the archipelago are being put into production. Brazil, Zaire, Indonesia, and many other tropical areas are said to contain large amounts of nickeliferous laterites which so far have been uneconomic to mine. The only operating nickel mine in the United States is at Riddle, Oregon, where there is a deposit thought to be a laterite formed during an earlier geologic time.

A great deal of work has been done and is still being done in an effort to improve the technology and treatment of these abundant nickeliferous laterites. They represent a very large resource of nickel, and it seems to be only a question of time until they are mined and used. The numerous large discoveries and the many properties recently put into production seriously depressed the price of nickel. This, combined with the slowing of the economy in 1970 and 1971, resulted in a surplus of nickel that will probably last for years.

Substitutes are available for almost all uses of nickel, but the prevailing low price, ready availability, and pleasing appearance of its alloys, along with extensive product research and excellent public sales relations, have made nickel a useful commodity in international trade. When Russia annexed Petsamo from Finland "for protective purposes" and took over the nickel deposits there which had belonged to the International Nickel Company, the Russians made a compensatory payment to the company, thus at least acknowledging the value of nickel.

Niobium

Niobium, or, as it used to be called, columbium, is an uncommon and to most people a comparatively little known metal. It is excellent for certain special steels, particularly in stainless steels to prevent intergranular corrosion, being of value in food canneries, for example, where without it the interstices of the steels would corrode as acids reached them. Niobium, when combined with carbon, forms an excellent stabilizer in steel. Such a stabilizer is necessary because a bridge, for instance, or the springs of an automobile undergoing almost continual movement will become granular; the constant movement produces fatigue, and the bridge or automobile springs will fail. Niobium reduces the effects of continued vibration, making a metal more secure.

Niobium also develops creep resistance where motion causes elongation of the grains of a metal or causes them to move over each other so that the metal becomes less stable.

In the past the ores of niobium came chiefly from Nigeria. They were sent to Norway, where cheap hydroelectric power was available, and were there reduced in electric furnaces in combination with iron to form ferroniobium.

A little over a decade ago niobium ores were discovered at Oka, in Quebec, Canada, and by 1964 about half of the world's production was coming from this district. Later large quantities of good ore were found at Araxá in Minas Gerais, Brazil. Brazilian deposits now supply approximately 60% of the world's demands. The increased availability of niobium following the discoveries in Canada and Brazil has led to expanded uses of the metal.

A small amount of niobium was formerly mined in Idaho, but the deposits were unable to compete with the better Nigerian ores. Essentially all niobium now used in the United States is imported.

Tantalum

Tantalum is another rare and slightly known metal, at present used as a shielding for thermocouples,* for immersion heaters in the chemical industries, and in a relatively new market in electronic capacitors.† It is recovered in small amounts from Zaire, Ruanda-Urundi, Malagasy, Mozambique, Portugal, and Brazil.

The available amount of this metal is so small that its extensive use has not been possible. If additional deposits are found, it almost certainly will become of great economic value both as an alloying metal and as a native element.

Titanium

Titanium is one more "modern" metal included in the ferro-alloy group. Titanium has roughly the same strength as steel and half the weight. Only a little is used in the iron and steel industries; in fact, in all steel plants titanium in ore is considered to be deleterious, though small amounts of titanium steels are available for special purposes, for which substitutes are generally possible. Far more important are titanium's uses in the form of the oxide as a stable white pigment and in its native state as one of the light metals.

Only since World War II has elemental titanium been available and used in appreciable quantities, most commonly where acid-resisting qualities are essential—for example, in the chemical and food-processing industries and in hardware for salt-water boats. It also provides the necessary quality of heat resistance when alloyed with aluminum and used in supersonic aircraft.

Most titanium metal is manufactured from the mineral *rutile* (TiO_2) but much of the raw material for pigment comes from the more refractory and difficult to treat *ilmenite* ($FeTiO_2$). Rutile is heavy and resistant to weathering, so it tends to concentrate in stream gravels and beach sands, from which it can be recovered. The best rutile-bearing sands are along the southeast coast of Australia and the coasts of Florida, and some rutile has been recovered from beaches along the west coast of Africa, in Sierra Leone. Large amounts of rutile were formerly obtained from the beaches of India and Brazil, but when these

* A thermocouple is used to measure temperature accurately. It consists of two dissimilar metals, joined so that a potential difference generated between the points of contact is a measure of the temperature difference.

† A capacitor is a device for accumulating and holding temporarily a charge of electricity.

sands were found to contain the radioactive element thorium along with rutile, both governments prohibited its mining and export. Both governments hope that thorium will eventually become valuable in the development of reactors for atomic energy, but in the meantime they have lost a badly needed market and source of foreign exchange.

Ilmenite is also a common constituent of sands—very few black sands do not contain at least a small percentage of this mineral. In addition, ilmenite is a common constituent of many ordinary rocks. Titanium is, in fact, one of the most abundant elements in the earth's crust, but because there are few deposits in which it occurs in economic quantities it is hard to obtain. Besides the beach sands, ilmenite is mined at Allard Lake, Canada, and Lake Sanford, New York.

Titanium ores, especially ilmenite, are extremely refractory and are very difficult to smelt, refine, and fabricate. Titanium melts at high temperatures (about 1820°C) and combines readily with oxygen at those temperatures; hence only with difficulty can it be retained in the metallic state while it is being worked.

Nevertheless, titanium, because of its abundance and wide distribution, should have a bright future, particularly in the transportation industries or other fields where strong, light, and noncorrosive materials are needed. International traffic in titanium and its minerals, although minor at present, could assume much greater value in the future as the technology and metallurgy of titanium become better understood.

Titanium illustrates several of the problems of mineral supplies. The amount available to industry has been quite small. However, if the use of supersonic jets expands beyond the military to commercial air travel, vastly larger amounts of titanium will be needed. Friction on the "skin" of supersonic transports causes temperatures high enough to melt aluminum. An alloy of aluminum and titanium that is approximately 40% titanium will overcome this difficulty, and in addition the strength of titanium will reinforce the air frames of the jets. These facts emphasize the importance of the discovery of new uses for the rarer and little known elements, the resultant problems of shortages, and the difficulty of finding substitutes to satisfy many vital needs.

Tungsten

Tungsten, also called wolfram, is one of the most familiar, most widely used, and most essential of all the ferrous elements. Heavy, hard, and tough, it is probably the best material known for the manufacture of hard carbides and hardened tool steels, and these uses have made it one of the most necessary of all metals. Tungsten carbide is used as the cutting edge in bits for drilling rock, principally in the search for oil, and in some circumstances can even substitute for diamonds in penetrating quite hard rocks. Tungsten is also used as metal powder or in x-ray tubes, in light-bulb filaments, and in alloy steels.

Substitutes are available for most uses of tungsten, but they seldom are as good, particularly in tool steel and drill bits.

Tungsten ores are mainly calcium and iron tungstates, oxides of calcium or iron and tungsten, which are sometimes combined with manganese and molybdenum. Tungsten is obtained mainly from Bolivia, Brazil, Canada, China, Korea, Portugal, Russia, Spain, Thailand, and the United States. In the past many small deposits were exploited in the United States under the incentive of high prices offered by the government, but only a few have been large and rich enough to compete successfully with foreign ores in a free market. Probably the largest and best deposit in the United States is near Pine Creek in east-central California. Considerable tungsten is also obtained as a byproduct from the molybdenum ores mined at Climax, Colorado.

Fig. 4-6. Tungsten production in 1970. "Others" include Australia, 3.6%; Brazil, 3.4%; Peru, 2.4%; Japan, 2%; Thailand, 2%; miscellaneous sources, 4.2%. Total production was 74,017,000 pounds. Source: United States Bureau of Mines.

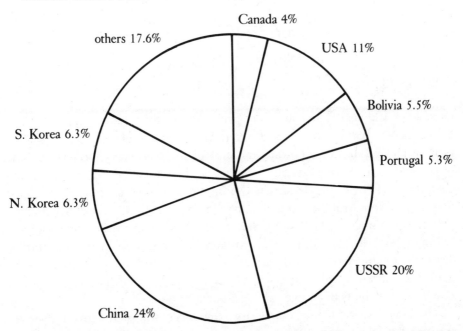

Canada 4%
others 17.6%
USA 11%
Bolivia 5.5%
S. Korea 6.3%
Portugal 5.3%
N. Korea 6.3%
USSR 20%
China 24%

Because Tungsten was in short supply during the early part of the Korean War, it was purchased in large amounts by the United States for storage in the

Fig. 4-7.   Tungsten consumption in 1970. "Others" include Czechoslovakia, 3.2%; China, 1.8%; Canada, 1.2%; miscellaneous sources,
4.5%. Total consumption was 83,603,000 pounds. Source: United
States Bureau of Mines.

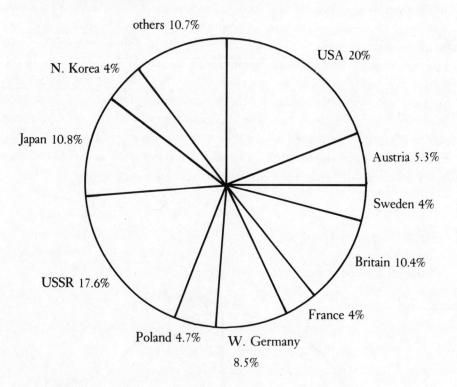

national stockpiles. The purchase price above the market appeared not
unreasonable at the start of the Korean War and stimulated a great deal of
activity and production. Toward the end of the war, the contract purchase
price was well above that paid on the world market. As a result, at the
completion of the stockpile purchase program, the price of tungsten dropped
drastically throughout the world and many mines had to be closed. At one time
all tungsten mines in the United States were idle except that at Pine Creek.
During 1961, Russia "dumped" tungsten in western Europe and further
disrupted an already disturbed market. Toward the end of 1965, however, the
market stabilized, surplus stocks were reduced, and prices began to advance
slowly so that the better mines are again able to operate at a profit.

The United States manufactures large amounts of tungsten steel but does not
produce enough tungsten ore to meet its needs because tungsten ore can be
imported at less cost than it can be mined here. In the past the United States
imported from China, which was, before the communist take-over, one of the

principal producers of tungsten in the world. China still has what may be the world's largest reserves and annually produces a surplus, some of which is sold in western Europe. When trade relations with China become normal, tungsten is probably one of the materials that the United States will import. Thus, we import tungsten from China; then we don't; then once again we do. Such a changing situation illustrates the effect on mineral industries of the complications of the international political economy.

Another example appears in the refusal of the United States, after World War II, to export tungsten steel or tool steel to Russia, which retaliated by cutting off the supply of manganese to the West. In this case both Russia and the United States found other sources of supply.

Tungsten is sold either as the metal or as ferrotungsten. The metal is commonly in powder form, 98.8% tungsten, and is sold in 1000-pound lots. Tungsten ore is sold by the short-ton unit of contained tungsten, or twenty pounds.

Vanadium

Vanadium is a ferrous metal that is well known and used in the manufacture of many alloys. Most vanadium goes into tool steels, high tensile strength steels, structural steels, and wear-resistant cast iron. Vanadium steels have made an effective armorplate on tanks and other vehicles because of their toughness and ability to withstand penetration, and vanadium steel provides some of our best cutlery.

The metal is a byproduct of the mining of many other metals and is trapped in the dust in smelter stacks, especially from the smelting of copper and from the uranium and phosphate industries. It has also been recovered as a residue from heavy oils—the dregs left in a shipload of Venezuelan petroleum after the oil has been pumped out are rich in vanadium.

During World War II Germany obtained an abundance of byproduct vanadium from the smelting of iron ores, many of which contain fractions of a percent of vanadium. At the present time no known mines are operating primarily for this metal, though formerly an excellent vanadium mine was operated at Minas Ragras, Peru, where a large body of asphaltic material contained a good percentage of vanadium.

The supply of vanadium is in general sufficient to fill the demand, principally for use in alloys. Many substitutes are available, but vanadium is considered preferable by most of its users.

Vanadium is commonly sold as the pentoxide or as ferrovanadium; its price varies depending upon the grade of the material.

Zirconium

Zirconium in the form of its silicate, zircon ($ZrSiO_4$), is commonly recovered from beach sands along with the titanium mineral rutile. About 80% of the zircon used in the world comes from the beach sands of eastern Australia. Florida beach sands also yield small amounts of byproduct zircon. Clear zircons, both colored and colorless, are known to many people as semiprecious stones with an unusually high index of refraction, which gives them the sparkle associated with gems. The metal zirconium is used to make ferrozirconium, a material which, when included in the steel-making process, behaves in much the same manner as manganese—it is an excellent though expensive scavenger that combines readily with oxygen. One of the outstanding properties of zirconium is its high permeability to slow-moving neutrons and its suitability for heat exchange units in nuclear reactors; the metal is finding additional uses as its metallurgy becomes better understood. There are many substitutes.

A common companion of zirconium in the mineral zircon is the element *hafnium,* which is seldom found without zirconium. Hafnium is in considerable demand in the atomic energy industries.

Other ferro-alloy elements

The rare alkaline earth metals cerium, lanthanum, and other similar elements that are at times included in the ferro-alloy group are all in scarce supply and expensive. They are useful mainly in small amounts for special purposes and in special alloys. Many of these metals are obtained from the Mountain Pass district in southern California.

Phosphorus, silicon, and boron, which may also be used as alloying elements in steels, are finding many uses where their peculiar properties are suitable. All three of these elements are obtained in the United States, where silicon and phosphorus are particularly abundant.

The future

The reader of this chapter, we hope, will by now have become aware, not only of the importance of the ferro-alloy metals, but of the changing conditions of their use. He will have noted that in general they are in limited supply and that the dubious political stability of their places of origin has affected economic and political relations between the industrialized nations on the one hand and, on the other, the underdeveloped countries that are comparatively rich in deposits.

Most industrialized nations have shortages of many of the ferro-alloy elements and must import them. One may assume that the automobile-buying public of these industrial nations will, however reluctantly, alter the shining symbol of their affluence and give up the chromium of their cars if this step becomes necessary. But what can take the place of manganese and tungsten in steel making, or of cobalt in the communications industries? What will be the political consequences of the need for these metals?

We may well worry about the future supplies of essential and irreplaceable metals. However, does it not seem probable that the uses of many ferro-alloy elements will be extended beyond the wildest dreams of today? The reader will have noticed that in numerous cases expanded uses have resulted from the discovery of new deposits, and that expanded uses would result in other cases if additional deposits could be found. He will have seen that expanded uses stem also from improved technology and from a better understanding of the metallurgy of some of the metals, as well as from their ever increasing role in atomic energy work, space and guided missile programs, jet aircrafts and electronics, and other industries that seem to point toward the future.

Consider the possibilities of the development of the nickeliferous laterites. Consider the increasing value, because of its noncorrosion, of stainless steel in marine equipment. As we look more and more toward the ocean for many purposes, including ocean mining, we must rely more and more on stainless steels.

The symbolism of steel to represent the civilization that all countries want for the future is not attributable to carbon steel alone. It is in large part due to the great variety of useful alloys of steel made possible by the ferro-alloy elements.

5

The Precious Metals

~~~~~~

We turn now from cheap, strong iron and steel that in their abundant quantities provide the basis of modern industrialization, and from the alloying elements that increase their usefulness, to a small group of rare, expensive metals. Instead of tons, we measure production in ounces; instead of skyscrapers, for instance, we discuss usage in photographic film, jewelry, and coins; instead of a price per pound of a few cents, we have a price per ounce of many dollars. Our emphasis shifts from the building of heavy industry to small units and to the medium of exchange upon which industry depends. Our subject encompasses the relationships of all the world's countries, for we are describing the basis of international trade.

The precious metals are gold, silver, and a group known as the platinum metals. Sometimes they are called the noble metals, partly because of their appearance and partly because, being chemically inert, they do not tarnish in pure air (although, as anyone who has to clean it knows, silver will turn black in the presence of sulfur-bearing gases).

The precious metals are today mined most often as byproducts of other mining, especially copper, lead, and zinc. This was not true in the past, when productive gold and silver mines left such names as the Mother Lode, the Comstock, Tonopah, and the Transvaal to mark history, but at the present time relatively few mines can be economically operated for any of the precious metals alone.

Each of the precious metals has its separate uses and all are highly valued. Gold and silver have a special importance because for many generations they have served as the basis of monetary systems; with, to a very minor extent, copper they are the only standard of value and the sole medium of international exchange that have ever been acceptable throughout the world. Therefore they have had a great influence upon economics, politics, and the history of man.

The precious metals are all scarce, yet their industrial uses are growing rapidly. A number of modern economists are attempting to de-emphasize the role of gold in international exchange with the claim that available amounts are insufficient to supply today's monetary needs. In spite of their efforts, the scarcities, and increasing industrial demands, gold and silver remain the world's medium of exchange, and economics and politics are still strongly influenced by them.

After discussing each of the precious metals, we discuss monetary metals in the political economy, considering current problems in relation to the price of gold and difficulties caused by the current shortages of both gold and silver.

Platinum metals

Platinum is one of a closely related group known collectively as the platinum metals because they are commonly found together in nature. The others are palladium, iridium, osmium, rhodium, and ruthenium. Each has unique properties and special uses in industry. All are needed but several are very rare. Platinum is the most abundant, the best known, and the most widely used.

Platinum, a steel-gray metal, is very heavy, 21.4 times as heavy as water. For all its rarity and nobility, it is not as popular for ornamental use as gold and silver, although it is used to some extent in jewelry. It looks considerably like silver—its name comes from the Spanish word for silver, *plata*—but differs from silver in being harder to work (it has a melting point of 1755°C while silver melts at 960°C) and in its lack of the warm patina and luster that silver acquires with use, for platinum never loses its cold, steely color. Neither has platinum been used as a backing for currency, being far too rare, though small amounts have been hoarded as possible hedges against inflation.

The value of platinum lies primarily in its industrial uses. Chemically it is so inert that vessels made of it will not combine with their contents, which is important to a chemist working with substances whose purity must be maintained. Its high melting point permits the chemist to heat a vessel's contents to high temperatures. Platinum also serves as a catalyst* in many chemical processes, for example the making of sulfuric acid and the refining of petroleum. Since platinum holds its shape when heated, nothing as satisfactory has been found for the making of modern synthetics such as nylon thread, in which a superheated mix is extruded from its platinum container through holes that must not wear but must maintain a uniform size. Additional amounts of platinum metals are used in the electrical industry and dentistry.

* A catalyst is an agent that causes a chemical reaction between two or more substances without being itself affected in any way.

New uses for the platinum metals are continually being found. In 1971 the Ford Motor Company announced that, in order to meet government-imposed standards on motor exhaust emissions, it would use platinum as a catalyzer in converters on cars sold in California beginning with the 1974 models, in order to reduce pollution by the exhaust gases. The company indicated that, if the California experiment is successful, all cars may carry similar converters in 1975. Where will this needed platinum come from?

Platinum in nature is found as the native element and is seldom combined chemically with other materials, although it is commonly alloyed with other members of the platinum group. It must be refined and purified before it is used in industry. It is found in many places, but only a few large deposits supply most of the world's needs. The largest are in Canada, the Union of South Africa, and the Ural Mountains of Russia. Smaller amounts are mined in Colombia. The United States has only the tiny deposits at Goodnews Bay, Alaska.

The world's production of platinum is entirely consumed by industry. Since the metal is so valuable and so chemically inert, stocks of it do not disappear. The problem of supply is simply that of finding and refining enough to serve the needs of an increasing population for new and growing industrial uses.

As with most metals, the pricing of platinum is cyclical; when the economy is depressed the demands for platinum decrease and the price falls. This was the situation in 1971, when the price dropped to its lowest in several years, $120 to $125 an ounce (some thousands of times the cost of steel).

The status of platinum in world trade seems unlikely to change appreciably in the near future, though its uses are certain to grow. This metal, like all the others of the platinum group, is an example of a mineral that fills important functions as no known substitute can, and its price shows what people will pay for something rare and indispensable.

Silver

Silver was known to the ancient Hebrews by a term meaning "pale," while the name given it by the ancient Greeks signifies "shine." Its appearance is familiar to us now, for it is the most common of the precious metals. Like gold, it has long been a symbol of desirability. In the sixteenth century Cervantes coined a proverb that has been repeated ever since, when he said, "Every man is not born with a silver spoon in his mouth."

In addition to its beauty, silver has characteristics that make it valuable. It is malleable and can easily be melted, worked, and molded; it is ductile, conducting electricity well; it is inexpensive relative to gold, though between 1965 and 1967 its price per ounce rose, varying from 90¢ to $1.775. Since that time it has fluctuated higher and lower; in 1970 and 1971 it sank to between $1.30 and $1.40 an ounce, but in 1972 it rose again, to $1.85. Early in 1973 the price passed $2 an ounce, and by the end of the year was more than $3. Early in 1974 the price passed $5 an ounce.

Fig. 5-1. New York silver price, monthly range. The variation over a six-year period shows the effect of supply and demand on price. It is also an example of one of the reasons why those who engage in the mining business need strong constitutions. Sources: *Investor's Reader* and Handy & Harman. Copyrighted and used with permission.

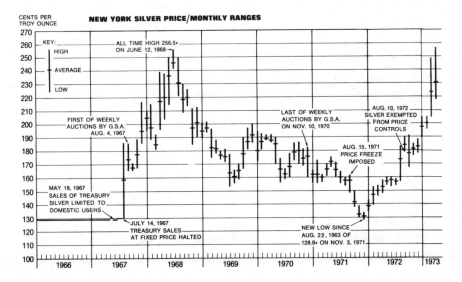

Silver is used primarily in the electronics industry, in the making of photographic film, and in silverware and jewelry; to a lesser extent it is used in coins. For centuries silver was the world's most common coinage metal. It has been ousted from this position mainly because its price increased so much in the past thirty years and because it is so badly needed in industry.

The superior electrical conductivity of silver has led to its use in the expanding electronics industry of this century, particularly where corrosion-resistant conductors are needed. While any single electronic unit may use only a minute amount, the increasing number of such units results in a large total consumption of the metal. In 1966 the use of silver in the electrical and electronics industries was estimated to have been about 36,000,000 ounces.

Its principal use continues to be in photography, where, although many substitutes have been proposed, no adequate substitute has yet appeared. From

31,000,000 ounces in 1964, 46,500,000 ounces of silver were utilized by photography in the United States in 1966, and 39,000,000 in 1972.

The consumption of silver in nuclear defense, solder, medicine, and chemistry, as well as in electronics and photography, has been expanding rapidly and will probably continue to grow. Other uses, such as in cassettes, could develop huge new markets. Already there is not enough silver for uses outside industry. Stainless steel is taking the place of silverware on the table, and less costly metals such as nickel and copper are replacing silver in coins. Yet even more silver is needed. Which uses are we to curtail? How are we to meet the increasing demands of industry even if we give up silver in coins and jewelry and silverware?

Silver is now obtained largely as a byproduct from the mining of copper, lead, and zinc, though a few districts, such as the Coeur d'Alene area in Idaho and some places in Mexico (Pachuca, Guanajuato, and others) and in Peru, have mines that operate primarily on silver ores. The scarcity of mines that produce mainly silver ores complicates the problem of increasing the production of the metal, since the demand for the metals with which silver is associated becomes a relevant factor. It is not always economic to increase a mine's production of copper or lead in order to obtain byproduct silver.

Formerly there were many silver mines, such as the famous Comstock Lode in Nevada, that mined principally silver ores with some gold. In these "bonanzas" the ores were near the surface; many were exposed. They were easily discovered by early prospectors and easily mined. Once they contributed large amounts of silver to the economy, but they have long been exhausted.

Silver is obtained in many areas of the world. Mexico and Peru are the leading producers, followed by Russia, Canada, and the United States. Producers of lesser amounts include Bolivia, Burma, Zaire, Argentina, Australia, Germany, Honduras, Japan, Korea, the Philippine Islands, Sweden, and the Union of South Africa. While new discoveries of silver ores have been made in recent years, with a few exceptions the discoveries are of little economic value and small, too small to prevent the development of future shortages. The silver market depends more and more upon byproduct silver from the smelting of copper, lead, and zinc ores.

The past few years have been difficult ones for silver producers in the United States. Finally, however, tight bureaucratic controls have had their shackles removed from silver, permitting a free market within which, in a few years, silver should become just another valuable commodity used in industry. The very large amounts of silver released when the United States abandoned silver coinage have nearly been absorbed by industry. The price of silver is expected to rise slowly until supply and demand reach a balance.

Fig. 5-2. Production and consumption of silver, in millions of ounces, in the free world, 1957–1970. The lower line shows production. The upper line shows consumption. The divergence of the two lines shows the deficit made up from previously mined stocks. Source: United States Bureau of Mines.

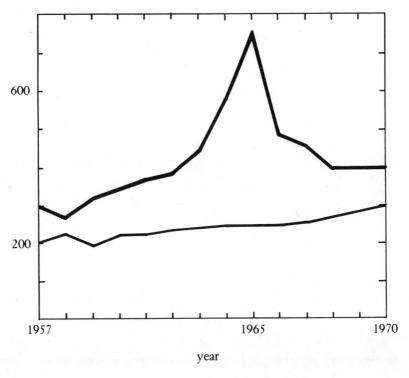

Gold

Only two metals have a color markedly different from gray—copper and gold. Both have intrigued mankind since ancient times, but the beautiful soft yellow color, the durability, and the scarcity of gold have made it especially attractive. Gold has been valued highly as far back as history is recorded. The founders of the very oldest civilizations fashioned gold ornaments that have been and still are from time to time dug out of the earth by archaeologists. In medieval times the alchemists tried to turn other metals into gold, which they considered the most perfect substance in the world. The gold treasures of the Indians in North and South America led to the conquest and colonization of Mexico and Peru, and the capture of the Spanish galleons bringing gold from the New World built up the wealth of England. A gold rush sparked the settlement of western North America. Throughout human history gold has

Fig. 5-3. Silver production in 1970. "Others" include Japan, 3.5%;
Bolivia, 2.2%; Sweden, 2%; East Germany, 1.5%; Honduras, 1.2%;
Morocco, 1.1%; Yugoslavia, 1.1%; miscellaneous sources, 9.8%. Total
production was 301,745,000 troy ounces. Source: United States Bureau of Mines.

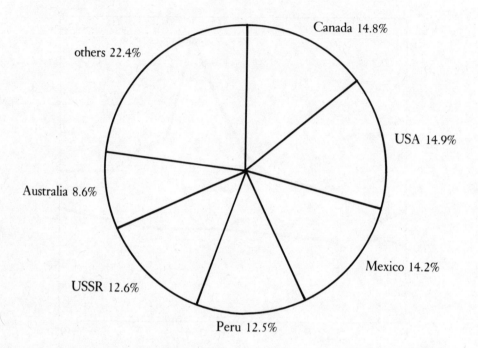

been a driving force, motivating nations as well as individuals to actions of
dramatic and far-reaching impact.

Why is this so? Why is gold important to men? What is there about this
particular metal that has made it more desired than any other?

The question cannot be answered satisfactorily by reference merely to the
properties of gold. By now, of course, its long tradition has surrounded gold
with an aura of glamour and worth, but how did this tradition evolve in so
many places through the centuries? Does the metal have an inherent beauty
that, enhanced by rarity, gives it a universal appeal?

All over the world, people guard and use gold jewelry, and a growing
number keep at least a few old gold coins. The meaning of gold to many people
is well illustrated by a man who came to the United States from Poland after
surviving two world wars. He insisted upon buying and keeping the best gold
jewelry for his wife, even though she did not especially enjoy wearing it. He
maintained that he had survived the two wars only because his family had

accumulated a considerable stock of gold coins which enabled them to purchase food, transportation, and eventual freedom. He considered gold as insurance against both inflation and national crises such as wars. He reasoned that gold could be sold when most other commodities would not be acceptable. Since in the United States people are prohibited by law from hoarding gold, the closest approximation possible is an accumulation of gold jewelry. With it this man felt ready for emergencies.

Another example of how people formerly valued gold is furnished by Sheik Ibn Saud of Arabia. When he first contracted to sell oil to the Arabian American Oil Company (Aramco), Ibn Saud specified that his royalties must be paid in gold; if dollars constituted the payment, the price was automatically increased one third.

Pure gold, defined as "1,000 fine" or 24 carat,* is so soft that it gradually wears away in jewelry, although in parts of the world, such as India, pure yellow gold is traditionally preferred to the harder alloys that commonly have slightly different colors. Gold may be alloyed with silver, copper, or other metals to increase its hardness and to form the white, red, or green gold generally sold in jewelry. In the United States "natural" or unprocessed gold could always be sold directly for use in jewelry and industry, but until 1968 so-called processed gold could legally be sold only through the government. "Natural" gold is gold that has not been subjected to recovery processes other than washing or panning; it forms an almost negligible part of gold production.

Fig. 5-4. United States gold production in troy ounces during the past century. Source: United States Bureau of Mines.

1880	. . . . . .	1,574,778	1940 4,862,979
1890	. . . . . .	1,589,018	1950 2,394,231
1900	. . . . . .	3,118,398	1960 1,679,800
1910	. . . . . .	4,657,018	1965 1,705,000
1920	. . . . . .	2,476,166	1970 1,743,322
1930	. . . . . .	2,138,724	

The principal use for gold has been as a backing for currency. Large amounts also go into ornaments and jewelry. Because of its corrosion-resistant

* Gold is measured by the troy weight system (named for Troyes, France) in England and the United States. Troy weight has 480 grains to the ounce, and 12 ounces (5,760 grains) to the pound, whereas the avoirdupois weight system, used for other commodities, has 437.5 grains to the ounce and 16 ounces (7,000 grains) to the pound. "Fineness" refers to the proportion of pure gold (or silver) out of a total of 1,000 parts. Gold is also measured by the carat, a unit of weight equalling a 24th part and used to express the fineness of a gold alloy, "14 carat" being 14 parts gold and 10 alloy.

properties, gold is finding increasing uses in industry, particularly in electronics, even though the limit on quantity, the price, high costs of discovery and recovery, and uncertainty about future supplies all tend to restrict new uses.

Gold is generally mined as the native metal, commonly alloyed with silver; it rarely combines with other elements. It has been produced in many places in the world. Numerous areas and even countries were settled as a direct result of the discovery of gold deposits. At present, the Union of South Africa, with its famous Witwatersrand (or simply "the Rand") district, is by far the largest producer in the world. Other leading producers are Australia, Brazil, Canada, Chile, Colombia, India, Peru, the Philippine Islands, Russia, the United States, and Venezuela. Many other areas yield minor amounts, for example the Fiji Islands, Japan, and several countries in Africa and Central America.

In the United States only one large gold mine, the Homestake in South Dakota, has managed to remain in production over a long period of time. Otherwise, gold production in this country has been largely a byproduct from the mining of other metals, particularly copper. In 1965 a property near Carlin, Nevada, came into production and a few years later the nearby Cortez mine was developed. These are open pit mines that generally contain between one third and one quarter ounce of gold per ton of ore; their development has increased production in the United States. Even so, in the United States the domestic market is approximately five times the domestic production.

Gold mining as an industry could be described only as distressed until recently. In many places in the world the mines have been kept open because of government subsidies. For a long time the prices of gold were tightly controlled and maintained at low levels; only in the past few years has a market been available except through the government and have governments allowed the price to seek its level for industrial uses. Still the market is far from free. In the United States (until 1974) and communist countries, people have not been allowed to own gold. The price increases allowed are only now beginning to equal the increases in costs of mining and refining that have occurred during the past years of inflation. The Homestake gold mine is operating at depths of as much as 7,100 feet; costs are rising rapidly, and this old property may well have difficulty in continuing its operations unless the price of gold is maintained in the future above $80 an ounce.

In production, the Union of South Africa is followed by Russia, with Canada in third place. The suspicion of large amounts of gold available in the Soviet coffers and the dominant position of an internationally unpopular government in South Africa have unquestionably influenced the pricing of gold and the world's monetary policies.

Fig. 5-5. Gold production in 1970. "Others" include United States, 3.8%; Australia, 1.3%; Ghana, 1.3%; Philippine Islands, 1.2%; Rhodesia, 1.1%; miscellaneous sources, 4.7%. Total production was 47,356,053 troy ounces. Source: United States Bureau of Mines.

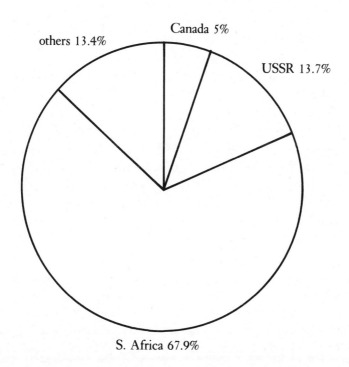

others 13.4%

Canada 5%

USSR 13.7%

S. Africa 67.9%

Monetary metals

The problems of monetary metals are being discussed everywhere by the world's economists, bankers, and politicians. Most of these problems involve "paper" gold or Special Drawing Rights (SDR's), the role of gold in the future, whether fixed currency rates should be re-established or currencies should be allowed to "float," and other subjects all complicated and abstruse. A few fundamental facts, however, are easy to understand.

• The world needs international trade in order to be prosperous and economically healthy. Anyone who has lived through a major war or read the history of one, or survived an extended economic depression, knows that when international trade is disrupted, local shortages result. No nation is entirely self-sufficient; all nations must trade.

• International trade in the world today requires a medium of exchange. To date only gold and, to a lesser extent, silver have furnished a medium of exchange that is acceptable everywhere.

• There is not enough gold or silver in the world today at the present prices to provide a medium of exchange and fill other important needs. The production of gold at the present prices is barely economic, and most of the world's production of silver is being channeled into industry.

Possibly economists and monetary experts will come up with an adequate and acceptable substitute for gold to back currencies in international transactions. Will everybody accept Special Drawing Rights? For how long and in what quantity? Are SDR's anything but fiat money, that is, mere promises that may not be worth their face value?

Before the gold standard was widely adopted, payment was usually by barter, the exchange of goods or services for goods or services.* A farmer paid for his groceries with sacks of wheat; Indians sold furs in exchange for beads, guns, and firewater; Marco Polo brought back from China gems and silks for which he had traded European articles. There are many examples of barter throughout history because it was a widespread practice extending to international trade, which, as it became more complex, was hampered by the exchange of items whose relative value in each transaction had to be established.

Today barter is still a general practice in a few remote areas of the world, and it is sometimes used for international exchange between countries of which at least one is industrialized. Russia, for instance, has given arms to Cuba and Egypt in return for mortgages on crops of Cuban sugar and Egyptian cotton. The intricate and large-scale business transactions of today, however, cannot be handled to any reasonable extent by the simple, clumsy methods of barter. There must be a medium of exchange that is usable everywhere if the trade necessary to modern society is to continue. The complications of barter that would be required to obtain large amounts of oil in Saudi Arabia, refine it at Aruba, and distribute and sell it throughout the world are almost unimaginable.

Over the years man everywhere has found three substances acceptable as a standard of value—copper, silver, and gold. Copper, though used in a few coins of lesser denominations like pennies and centavos, is not rare enough to be stable in price. Also, it is needed in modern industry, where it has many important uses. Its monetary use today is localized and negligible.

Silver as a standard is also inadequate, both in supplies and in value, in spite of its long history as a backing for currency.

From 1933 to 1963 the Treasury was the largest buyer of silver in the United States. It accumulated more than 3.5 billion ounces, of which about two billion ounces were issued as coins. In January, 1967, the United States

* There were some early media of exchange—Indian wampum, for example.

government passed legislation to remove silver from coinage, and as a result the Treasury was able to sell its surplus silver at weekly auctions beginning in August of that year. These sales brought in more than five hundred fifty million dollars, the prices ranging from $2.43 an ounce for 0.996 fineness in May, 1968, to $1.50 an ounce for 0.900 fineness thirteen months later. As the monetary value of the silver was only $1.29 an ounce, the Treasury made a profit.

On November 10, 1970, the United States Treasury sold the last of its surplus silver stocks and went out of the silver business for the first time in 194 years. At that time the government retained about 165 million ounces in the Defense Department's strategic stockpile and an additional 35 million ounces held for minting an Eisenhower 40% silver dollar. This dollar was designed for collectors; the supply was quickly bought up at $10 each when it appeared late in 1971.

World demand for silver has been rising about 2% a year, to approximately 365 million ounces in 1970. At that time new production accounted for only about 250 million ounces a year, according to the Silver Users Association in Washington, D.C. An additional 60 to 80 million ounces a year are recovered from scrap materials, principally old film. This leaves an annual minimum deficit of 35 to 50 million ounces that is now supplied from decreasing inventories.

Immediately after the end of United States Treasury sales, the market dropped sharply (by about 22¢ an ounce). This was commonly ascribed to the fact that silver users had been anticipating a shortage after government sales stopped and had protected their future needs by building large inventories, thus depressing the market for subsequent sales. The major silver consumers used silver from these large inventories when the price rose, thereby exerting great pressure to keep the price down. Then, too, large numbers of coins were hoarded in the expectation of future high prices, which were just beginning to materialize in 1973. This hoarded silver is still in part hanging over the market and only slowly being reduced in quantity; it has a depressing influence on price. As this hoard is reduced silver prices should gradually rise.

Even if its price should rise to unexpected heights, silver will continue to be, because of the expansion of the photographic and electronic industries, of more value to industry than as a medium of exchange or currency backing.

Since both copper and silver are needed in industry and cannot be used as a universal standard of value, gold remains the only substance that is acceptable everywhere. It sustains the modern monetary systems. Trade all over the world depends upon it. As its supply becomes increasingly inadequate, the foundation of the world's political economy becomes insecure and must be strengthened. How can this be done with a minimum disruption of trade?

The reader cannot be expected to answer such a question at this point. Indeed, he must have information as a necessary background for even some understanding of what is involved in finding an answer. To provide that background, we consider next some facts about the use of gold in monetary reserves.

Since the discovery of the New World and its treasures, world stocks of processed gold amassed by man are estimated to have reached by the end of 1966 a total of about 77,000 metric tons. Of this amount, 14,000 metric tons (18%) are estimated to have gone into jewelry and industry, 17,000 metric tons (22%) into private hoards, and 46,000 metric tons (60%) into monetary reserves. How big are 46,000 metric tons? How great a quantity is this gold which has lain in Fort Knox and the vaults of other nations? Some people envision large fortresses filled with the metal. Actually 46,000 metric tons will cover a tennis court (36 feet by 78 feet) to a depth of about 30 feet. This is the gold that in the past has backed the currency of the world.

The amount of gold that was adequate for world currency reserves in 1900 is clearly not adequate in the 1970's, with their greater population and the expansion of international trade that has taken place during the intervening years.

When a mineral is in demand and supplies are scarce, the first solution proposed is a substitute. But gold is a political and economic symbol throughout the world. How can a substitute be found for a symbol acceptable to all people? It cannot be found in a laboratory by chemists or physicists. It can be found only by political economists. That they have not succeeded to date is a cause of current troubles in the international economy.

If gold is not used in international trade, what can be? Would one nation accept another's promise to pay? Would Canada, for example, have taken Russian rubles for the wheat it sent to Russia, for which Russia paid in gold?

The price of gold for several years was based upon a two-tier system, one price for monetary gold and another price for the free market. This system could be maintained so long as the two prices were close together, but as the free market price gradually increased so did the difficulties with currencies. It is fortunate that the system was abandoned, because the wide divergence of value of the two tiers tempted some nations to sell monetary gold on the free market to make a profit. Yet the fact that none of the leading currencies is now tied directly to gold creates other problems. Discipline in the money market is inadequate without the restrictions imposed by a standard such as gold. Without a standard, money managers in a government are able to issue freely paper money. Such a currency quickly depreciates in value, which means inflation in that nation; and when there are many floating currencies, inflation is worldwide.

At the Bretton Woods conference, held in 1944 by many of the nations that now form the International Monetary Fund, the United States agreed to buy or sell gold at $35 a troy ounce, plus a commission of ¼% for handling, which meant that dollars could be used in international dealings and could be turned in for gold at any time. This situation continued until the summer of 1971, when inflation, the rigidity of the monetary system, and gold shortages caused the United States to suspend all payments of gold for dollars. A continuing unfavorable balance of trade was fast draining gold from the Treasury, and clearly the United States had been living well beyond its capabilities to pay. Currency revaluation was necessary. The United States therefore followed the wishes of other nations and devalued its currency by 8% in terms of gold, while other nations revalued their currencies upward. But can any fixed currency ratios be more than temporary expedients?

Continued worry about the value of the dollar, which was backed only by the promise to pay in goods or commodities, caused a selling wave throughout the world early in February of 1973. Nations were unable to retain the ratio of their currencies to the dollar, even though the West German government bought as much as six billion dollars in six days. Around the middle of February the dollar was further devalued, this time by 10%. But the basic problems still existed. No acceptable standard of value backed the dollar. The United States continued through 1973 to export more dollars than were returned. Inflation persisted; hence people in other countries thought that the future value of the dollar would be less than the present value and they did not want to hold dollars.

With gold at a fixed price, whether $35 or $45 an ounce, or even at a free market price of as much as $100 an ounce, many gold mines are today not economic to operate, because of the increased costs of mining. Some mines that are marginal are maintained in operation only because of government subsidies or other form of financial relief. No government aid is given gold mines in the United States, and profits from gold mining here have decreased to such an extent that most mines have closed. Even so famous a district as the Mother Lode of California, if it could return to the height of its production and prosperity, would be unprofitable to mine at a monetary price of about $100 an ounce and the current cost of extraction. The development of new gold mines is no longer a major factor in the mining industry, and most mining companies do little or no exploration for gold. The slowly increasing demands of industry can raise the price only over a period of years. If gold mining is to continue, the price of gold must increase at a rate equal to the increase of the prices of other commodities.

The problem of gold is not a problem of mineral production. It is an economic and psychological and political problem, complex and confusing. It is compounded by the urgent needs, mounting rapidly, of international trade, which means credit, which means payment in acceptable currency, which means currency based on some international standard of value.

The future of the precious metals

The more one studies the dilemma of the precious metals, the more one sees the ramifications of international trade, involving not only the interests of governments of all nations, but even the individual's instinct for self-preservation. Here are problems whose solution needs the kind of universal cooperation that has been envisioned by those who foresee "one world" for the future. Here again, in the case of silver, we are brought to the conclusion that the problems of mineral shortages are growing. They have not been solved and they must be solved if our civilization is to continue.

Solutions to the complex problems of gold that have not yet been solved by the world's leading political economists cannot be offered in this book. One point, however, can be made with confidence. So long as people want gold or put faith in it as a medium of exchange, and so long as the price is controlled and set at a figure that discourages and in most cases prevents its mining, gold will be in increasingly short supply.

What will happen to it if a new unit like Special Drawing Rights takes its place? Will it become only another minor commodity, with increasing uses in industry and a modest price that fluctuates on the free market? Will man's attitude toward this metal change? Or will his imagination, his mind, and his actions continue to be influenced by the power of gold?

6

The Nonferrous Metals

~~~~~~

The nonferrous metals have this name because they are not used as alloys with iron. They fall naturally into two groups, the base metals and the light metals. The base metals, so called in contrast to the noble or precious metals, include copper, lead, zinc, tin, and mercury. The light metals include aluminum, magnesium, beryllium, lithium, and titanium.

Several less well known nonferrous metals are used in minor amounts. Cadmium is obtained as a byproduct from the mining of zinc, and bismuth and germanium are both byproducts from the mining of other metals. Each of these elements has a few uses for which it is pre-eminent. Cadmium forms a bright canary-yellow pigment, bismuth is used in alloys that melt at low temperatures, and germanium is useful in the electronics industry.

Because most of the nonferrous metals are available only in ores of low grades, their mining requires large amounts of venture capital.

The base metals

Copper. Copper is probably the best known of the base metals. Its warm glowing color in both ornamental and utilitarian objects, such as bowls and kettles, brightens many a home in the industrialized nations, and the economies of some of the less developed countries depend entirely upon it.

This metal possesses qualities that make it useful in many industries. It is an excellent conductor of heat as well as of electricity, it wears well, and it is malleable, ductile, and reasonably inert. About 60% of the copper sold goes into the electrical industry. The construction and chemical industries require it for a number of uses. Copper is also used in low-denomination coins, in kitchenware, and as an ingredient of bronze, brass, and other popular alloys.

Because copper is relatively expensive, substitutes, especially aluminum, often take its place. For example, aluminum increasingly provides the cable for power lines, and copper pipes are used in plumbing only when the purchaser is willing to pay more in order to get pipes that will last longer. Copper in large amounts can no longer be afforded for use in buildings, as it once was for roofing, drainpipes, and gutters.

Another limitation on the use of copper is the fact that it will slowly oxidize, turning green in the process; therefore silver and, at times, gold are preferred in the electronics industry, where oxidation cannot be tolerated.

Copper is mined principally in the United States, Chile and Peru, Zaire, Canada, Australia, and Zambia. Lesser amounts are obtained from Japan, the Philippine Islands, Russia, Sweden, and Yugoslavia. Gold, silver, and molybdenum are byproducts of copper mining, increasing the value of the ore and making its mining more economically feasible.

Approximately two thirds of the copper produced in the world come from four nations—Chile, Zambia, Zaire, and Peru—which have banded together to form the Council of Copper-exporting Countries, commonly called CIPEC from its French initials. Zambia relies upon copper for about 90% of its export earnings, Chile for about 70%, Zaire about 50%, and Peru about 20%. These four countries account for about 80% of the total primary copper sold in the export markets, and the price of copper is vital to all of them. The stated purpose of CIPEC is to stabilize the market for copper and to assure the producers of a steady, profitable trade. However, CIPEC operates very much like an old-fashioned cartel. In spite of the essentiality of a steady market to the countries concerned, the primary objective of CIPEC appears to be to assure them of ever more revenue.

CIPEC has proposed various means of keeping the price of copper up and steady, but so far the countries have failed to reach any satisfactory agreement as to how they can do so. Four methods of price support are open to them.

1. Cutbacks in production.

2. The establishment of fixed prices.

3. Regulation of exports.

4. Purchasing surplus copper and storing it to create a scarcity in order to maintain an artificially high price.

As *Metals Week* pointed out (March 1, 1971), methods 1 and 3 involve social problems and would mean a loss of jobs that the relatively weak governments of the exporting countries probably cannot endure. No. 2 would require limitation of planned expansion to which the nations are strongly committed; this would mean holding the present line while the cheaper stocks now in inventory were reduced. No. 4 is probably the best choice, but it calls for more cooperation among these producers than has so far been evident.

Fig. 6-1. Copper production in 1970. "Other Europe" includes Yugoslavia, 1.4%; Poland, 1.1%; miscellaneous sources, 2.1%. "Others" include Australia, 3%; Philippine Islands, 2.4%; South Africa, 2.4%; Japan, 2%; China, 1.1%; Mexico, 1%; miscellaneous sources, 4.3%. Total production was 6,566,643 short tons. Source: United States Bureau of Mines.

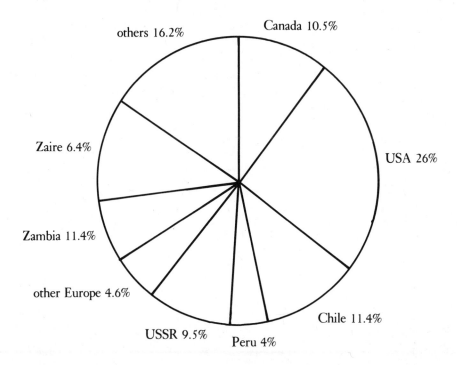

The need for copper has been growing at a very rapid rate throughout the world. One effect of CIPEC activities has been to encourage intensive exploration in areas other than its four member nations, an effort that continues to meet with considerable success. Large new deposits of copper ores have recently been found in Australia, Bougainville, Papua, the Philippine Islands, Puerto Rico, and many other places. Exploration is proceeding, although it has been curtailed by the economic slow-down of 1970 and 1971 and by recent expropriations. Some nations, particularly Canada and Australia, have surplus copper for export but so far have refused to join CIPEC.

As with almost all mineral commodities, many regions suffer from a deficiency. Europe has little copper and must depend primarily upon Africa. Japan, which produces some copper and was once one of the leading suppliers in the world, now imports large amounts from the Philippine Islands, Australia, and South America.

The United States is nearly self-sufficient in copper; some of its deposits, however, are low grade, yielding only seven to ten pounds of copper for each ton of ore. This means that the costs of extraction and refining are high, and mining is economically feasible only with modern mechanical equipment and inexpensive energy.

Fig. 6-2. Production and consumption of copper, in thousands of short tons, in the United States, 1900–1970. The dotted line shows production. The solid line shows consumption. In this case the divergence of the lines changes at approximately 1920. Up to that time, the nation produced more than it consumed and exported copper. After that time, the nation produced less than it consumed and imported copper. Source: United States Bureau of Mines.

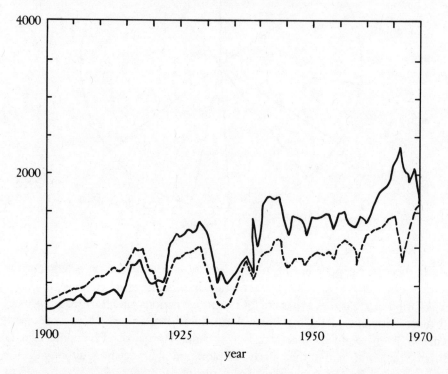

The price of copper is an interesting economic and political study in itself. The availability of substitutes has influenced many copper-producing companies to attempt to hold down prices in order to remain competitive and retain their markets. Generally, when a market such as the cables on transmission lines is taken over by another metal such as aluminum, that market is lost to copper. The policies of independent copper companies, in their efforts to keep prices down to retain markets, have run into direct conflict with the stated policies of the CIPEC countries that are trying to increase prices and thereby their incomes. Thus prices fluctuate. The London Metals Exchange, upon which CIPEC prices are based, issues daily quotations.

Since in the past the price of copper has been notoriously cyclical, in spite of all efforts to stabilize the market, estimates of price in the future are difficult to make. Toward the end of 1970 and early in 1971, a surplus of copper began to accumulate in world inventories. The prices of the metal started to go down. Large new deposits were scheduled to begin production, industrial demands slackened, and military demands for Vietnam decreased. On the other hand, the expropriation of the copper mines in Chile, the inability of the Chilean government to operate the mines so as to meet outstanding contractual obligations for delivery, an economic upturn, delays in opening several mines, and strikes at several others all served to exert upward pressure on the price of copper.

When Rhodesia closed the border with Zambia, it claimed that Zambia was protecting guerrilla activities. Although the Rhodesian government said it would permit the export of Zambian copper to the seaboard, nevertheless conditions were such that Zambia refused and retaliated by closing its border. Zambia has persisted in this closure, even though Rhodesia has agreed to reopen the country to Zambian materials. Transportation of Zambian copper by routes other than through Rhodesia is slow and inadequate. As a result, production of Zambian copper has declined considerably. This, combined with the other factors and an improved industrial outlook early in 1973, has led to an increase in the price of copper.

Lead and zinc. Lead and zinc ores, commonly found together, are widely distributed in the world.

Both are of great economic value, being utilized by many industries. Lead is used in gasoline, ammunition, storage batteries, pigments, sheathing for underground electrical cables, and numerous alloys. It is toxic to humans and therefore is now used far less than formerly in pigments and glazes. Zinc oxidizes very slowly and as a coating affords protection. Such coating is called galvanizing and is commonly seen on items like buckets and sheet or corrugated iron. Zinc is also used in vulcanizing rubber, in dry cell batteries, pigments, die castings, and in bronze, brass, and other alloys.

The use of lead in gasoline has been the subject of heated debate as people concerned about clean air have insisted that the lead be removed. Lead additives in gasoline, which account for about 20% of the lead market, are known to provide anti-knock qualities that make fuels smoother burning, more efficient, and less damaging to engines, but lead in large quantities is poured out of

Fig. 6-3. Production and consumption of lead, in thousands of short tons, in the United States, 1900–1970. The dotted line shows production. The solid line shows consumption. During this century the United States has changed from an exporting nation to an importing nation. Source: United States Bureau of Mines.

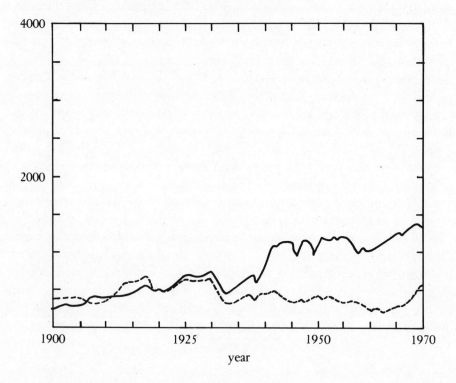

exhausts into the air, from which it quickly settles onto the ground. This is the lead that people concerned with the environment want taken out. Many statements, based only in part upon adequate knowledge, have been made concerning both environmental dangers resulting from the lead and those that would result from the removal of the lead. The over-all result of the controversy has been a trend toward removal, and most oil companies now market nonleaded or low-leaded gasolines; probably the total removal of lead from gasoline is only a question of time. Sudden complete removal, as advocated by several organizations, should be avoided—it would cause serious difficulties. In addition to economic problems such as the necessary and expensive restructuring of refineries, much is yet unknown concerning the proposed substitutes for lead such as benzene, toluene, and xylene. Emissions resulting from the use of these substitutes may be worse than the emissions resulting from the use of lead, since the former are likely to increase the aromatic compounds known to contribute to smog and eye irritation, some of which are considered to be carcinogenic.

Substitutes exist for most, but not all, of the uses of both lead and zinc.

Since nature has been unusually generous in distributing them, the ores of both these metals are found in many countries. The United States has sizeable supplies available within its boundaries, although it is now a net importer of both. Zinc is in somewhat shorter supply here than lead, but enough of both is available for some years to come.

Fig. 6-4.   Production and consumption of zinc, in thousands of short tons, in the United States, 1900–1970. The dotted line shows production. The solid line shows consumption. During this century the United States has changed from an exporting nation to an importing nation. Source: United States Bureau of Mines.

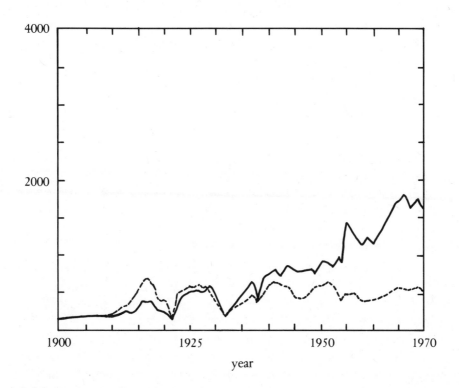

Australia, Canada, Mexico, Peru, Russia, and Rhodesia are large producers, and all except Russia depend upon the export of these metals for much of their foreign exchange. Russia has large reserves in the Caucasus, Abkhazia, West Siberia, East Siberia, the Ural Mountains, and Kazakstan. Several other

Fig. 6-5. Lead production in 1970. "Other Europe" includes Bulgaria, 3.5%; Sweden, 2.4%; Spain, 2.1%; Norway, 2%; Ireland. 1.7%; Rumania, 1.3%; West Germany, 1.2%; France, 1%; and Italy, 1%. China, 3%; Japan, 2%; Morocco, 2%; North Korea, 2%; Southwest Africa, 2%. "Others" include Argentina, 1%, and miscellaneous sources, 5.6%. Total production was 3,750,826 short tons. Source: United States Bureau of Mines.

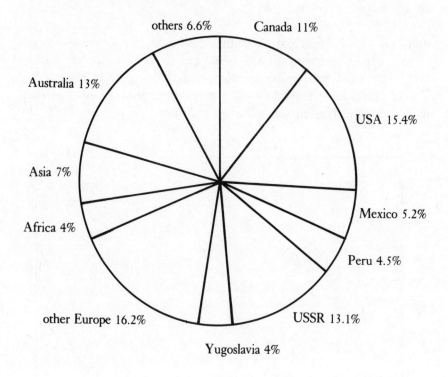

countries produce smaller amounts for domestic consumption or export; they include Argentina, Bolivia, Burma, China, Great Britain, Guatemala, Italy, Japan, Morocco, Poland, Portugal, Southwest Africa, Spain, Sweden, Tanzania, and Yugoslavia.

The extensive distribution of these metals means that if the price is raised only a little, an increase in production is immediately apparent, and conversely when the price is lowered slightly, production is almost immediately cut back. This situation tends to stabilize price, and, while lead and zinc fluctuate normally, their gyrations are far more restrained than those of copper.

The pricing of lead and zinc in the United States is different from that of most metals. Until recently it has been customary to quote prices that vary, depending upon the location, but based upon f.o.b. St. Louis, Missouri. Now,

Fig. 6-6. Zinc production in 1970. "Other Europe" includes Poland, 3.4%; West Germany, 2.2%; Italy, 2%; Yugoslavia, 1.8%; Ireland, 1.7%; Spain, 1.7%; Sweden, 1.7%; Bulgaria, 1.6%; Finland, 1.1%. "Other Asia" includes North Korea, 2.3%; China, 1.8%; Iran, 1%. "Africa" includes Zaire, 1.9%; Southwest Africa, 1.2%; Zambia, 1%. Total production was 6,060,604 short tons. Source: United States Bureau of Mines.

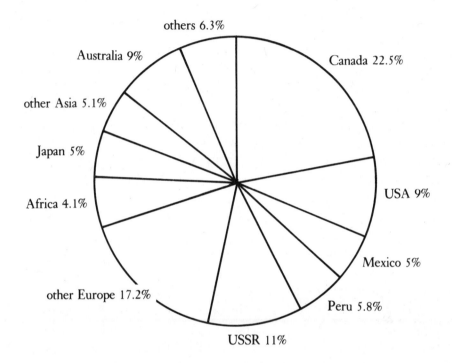

however, prices are quoted at one level throughout the country. Discounting has been common practice, which means that the quoted price signifies little in terms of actual selling price. If the price is 18¢ a pound, for instance, a seller may let a purchaser have a supply of metal for 16¢ a pound; such selling at a reduced price is called discounting. It is influenced by supply and demand and by the quantity, since the larger the bulk of a purchase, the greater the discount is likely to be. The recent effort to discontinue discounting has so far been only in part successful.

The prices of lead and zinc in the United States were in the past ordinarily two or three cents a pound higher than elsewhere, which encouraged foreign producers to sell here, thus flooding the market. As a consequence, our government at various times imposed duties and quotas for the protection of domestic industry. When world prices are competitive, quotas may be abolished.

Tin. Tin is an element that has been sought since ancient times, when the deposits of Cornwall and Spain were exploited by the Phoenicians and Romans.

Minor amounts are still obtained from these old mining sites, but by far the major quantity of tin comes from a small number of large producers in only a few areas—Asia, principally the southeast (Malaysia, Indonesia, Thailand, and China), Africa (particularly northern Nigeria and Zaire), and Bolivia. Several of the governments of these unindustrialized countries depend to a large extent upon the sale of tin for foreign exchange; the sale of tin at a fair price and reasonable stability of the market are essential to their welfare.

Very little tin is mined near the large industrial centers of the world. Brazil claims that a large deposit of tin ore has been discovered in the far western states of Acre and Rondonia, and the Renison-Bell mine in northwest Tasmania has become a large producer. In spite of these discoveries, the metal has been in short supply and the price has been rising. It is therefore understandable that the small number of producers, their locations, and the political instability of their governments have caused great concern among the users of the metal and have encouraged exploration in many places, such as Australia, where small amounts of tin are widely distributed.

The reasonably effective cartel known as the International Tin Committee has been organized in order to stabilize the price. This cartel maintains a buffer stock of tin that is used to control price and market practices. Quotas for producers are established to assure the producing countries fair income. This cartel is supported by both producers and consumers with the exception of the United States, which traditionally does not favor cartels, but the United States does send observers to meetings of the group and has not actively opposed its formation. The International Tin Committee continues to exert a steadying influence upon what was formerly a fickle and unstable market. Thus it benefits not only buyers but countries like Malaysia and Bolivia where tin is the principal item of export and the maintenance of a steady price structure, steady markets, and reasonable profits are necessary for routine government fiscal operations. The price of tin on the world market is generally quoted in British pounds per long ton of metal.

Tin is used as a stable coating over thin sheets of iron—in tin cans, for instance, which actually contain a negligible amount of tin but in most cases would be valueless without its coating, which ensures that the cans may easily be sealed hermetically with heat.

Tin is also used in the manufacture of pewter, bronze, and other alloys.

When pewter was originated, the best was an alloy of tin, copper, zinc, and antimony while the inferior contained lead. Eventually, because of its relatively high cost, the proportion of tin decreased while the proportion of the cheaper ingredient, lead, increased. When some cases of lead poisoning were traced to the use of pewter pitchers, the alloy fell into disfavor. Recently its popularity has revived, but modern pewter once again is an alloy of zinc, copper and tin.

Fig. 6-7. Tin production in 1970. "Africa" includes Nigeria, 3.4%, and Zaire, 3%. "Others" include Brazil, 1.3%, and miscellaneous sources, 5.4%. Total production was 226,569 long tons. Source: United States Bureau of Mines.

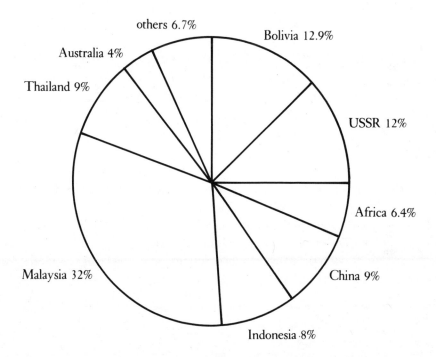

Mercury. Mercury, another essential material, is the only common metal that is liquid at ordinary room temperatures and can therefore be used in thermometers, electrical thermostats, barometers, and other temperature and pressure control equipment. It is also an excellent electrical conductor. Mercury has many uses—in explosives and fulminates, in the small button-size batteries of hearing aids and walkie-talkie equipment, and for numerous additional industrial and manufacturing purposes. An increasing quantity is needed for modern electronics, where, although the amount required by any single instrument may be small, the number of instruments requiring it is large.

Mercury has been used in agriculture to coat seeds in order to suppress fungi, but after some unfortunate occurrences when people ate the treated grain seeds, even though they were stained red, the practice has been terminated.

The paper-making and caustics-manufacturing industries, both of which formerly used large amounts, have been forced to decrease their use of mercury as it has come under increasing pressure from environmentalists. Although the claims of mercury poisoning have not yet been substantiated as a widespread danger, stringent regulations on the use of mercury are forcing those industries to change their processes in order to eliminate the dumping of large quantities of mercury-containing wastes.

Mercury comes mainly from three districts in Europe: Almadén in Spain, Idria in Yugoslavia, and Monte Amiata in Italy. Lesser amounts are produced in the United States (Alaska, and some western states), China, Mexico, and a few other countries.

The metal is usually mined as the mineral cinnabar, which is mercury sulfide. Cinnabar is a bright red color and in large pieces forms a highly prized semiprecious stone. Mercury in the liquid form is found in a few mines and oil wells.

Mercury has for generations been sold by the flask, a bottle-shaped container that holds seventy-six pounds. Efforts to change the container to a more convenient shape have so far failed. The price is usually quoted by the flask. Over the years, the price of mercury has been as mercurial as the metal itself; no other metal has had greater price fluctuations. In the past few years the price has been considerably depressed because so many industries have abandoned the use of mercury on account of proclaimed health hazards.

In spite of its potential dangers, mercury remains a metal that is essential, has no substitute, and is in limited supply.

The light metals

Aluminum. Aluminum has a special value in the construction, electrical, and transportation industries. It is used in all kinds of structures where a combination of lightness and strength is needed and where a lack of rigidity is acceptable—aluminum bends easily. It cannot be used for the main support of a skyscraper or bridge, for instance, since if it were, the bridge or building would sway, but its flexibility makes it pre-eminently suitable for the construction of airplanes. In fact, the convenient availability of aluminum, its light weight, its reasonable cost, and the fact that it can be readily fabricated into shapes have made it ideal for the manufacture of airplanes and other transportation vehicles.

Aluminum is offering serious competition in many fields to older, well established materials—in the development of corrugated sheets, cans and containers, and cooking utensils, to mention only three. Aluminum cable, because of its light weight and low cost, has largely replaced copper cable in electric power transmission lines, but aluminum is not as good a conductor as copper and therefore to transport the same amount of current requires larger wire. Aluminum foil and sheets are finding wide application in homes and in the building industry.

About 90% of all aluminum ores is used in the manufacture of aluminum metal; the other 10% goes into alumina, the aluminum oxide that is one of the widely used abrasives. Alumina, very refractory and insoluble, is also used in the chemical and refractory industries.

Aluminum in powder form will burn fiercely under certain conditions, such as abundance of oxygen and high temperatures.

Since aluminum metal is stable in the air under ordinary conditions, items such as old containers and foil tend to accumulate and do not corrode or disintegrate as do those made of iron. Several aluminum companies have begun active recycling campaigns in efforts to collect and use scrap aluminum and to avoid large quantities of unsightly refuse.

Aluminum is one of the most abundant elements in the earth's crust, and concentrations of its ores are spread over the world. The best known economic deposits are in Guyana (formerly British Guiana), French Guiana, China, France, Hungary, India, Jamaica (which is now said to be one of the largest producers in the world), Surinam, and Yugoslavia. Lesser amounts come from Brazil, Russia, and the United States. The large, newly discovered, and conveniently situated deposits of Australia now being developed are certain to have a profound influence on the aluminum industry. Sizeable undeveloped reserves are reported in Ghana, Angola, and elsewhere in tropical Africa, with smaller reserves existing in several other places as well. Because the ores of aluminum are oxides (bauxite and boemite) that are insoluble under ordinary conditions of weathering, they have accumulated especially in tropical areas where weathering is extreme and more soluble materials are leached or washed away. There the ores were formed by tropical weathering of rocks that contained large amounts of the aluminum-rich minerals.

In addition to the aluminum oxide ores, very large quantities of aluminum-rich clays exist in many areas.

The manufacture of aluminum metal from its ores requires huge amounts of electrical energy. Therefore aluminum smelters have been constructed in places such as the west coast of Norway and Kittimat in western British Columbia, Canada, where there is no aluminum ore but where there are large amounts of cheap hydroelectricity. The potential of tropical Africa is tremendous, not only because of its partly known and still undeveloped ores,

Fig. 6-8. Aluminum production in 1970. "Other Europe" includes France, 3.9%; West Germany, 3.2%; The Netherlands, 1.5%; Spain, 1.2%; Poland, 1%; Rumania, 1%; Switzerland, 1%. "Others" include Australia, 2.1%; India, 1.7%; China, 1.3%; Ghana, 1.2%; miscellaneous sources, 9.9%. Total production was 10,655,000 short tons. Source: United States Bureau of Mines.

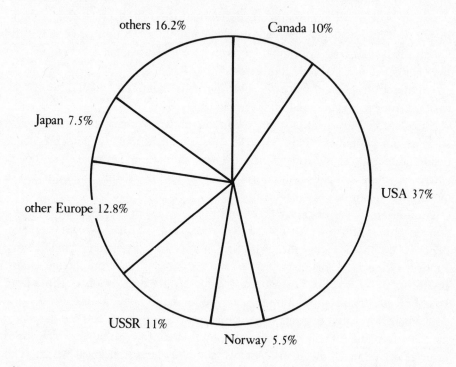

but also because of the possibility of harnessing the power of the Congo and other African rivers. With the completion of the Volta dam, aluminum production will begin in Ghana.

Publicly financed power dams and generation and transmission of electric power constitute a form of public subsidy for the aluminum industry. Such subsidy is seldom available to competing industries and metals. As energy becomes more expensive and is needed in other areas of the economy, aluminum may start to lose some of its market, especially where other metals that require less energy to fabricate are available. A bottle or a "tin" can may need less energy to make than an aluminum can.

Since reserves in the United States fall far short of the amount needed, most aluminum ores are imported here from Jamaica and Surinam. As energy becomes scarcer and higher priced in the United States, this nation may have to turn more and more toward imported aluminum bars, or ingots. Large deposits of aluminum-rich clays are known in the United States; they can be mined cheaply but require more energy to reduce to the metallic state than do the ordinary oxide ores.

It is a good thing for mankind everywhere that tremendous reserves of aluminous materials exist in various places and can be mined at low cost. Given only the plentiful supply of energy that will be needed to reduce the ore to the metal, aluminum should be available for many years.

Aluminum is sold by the pound in the form of thirty-pound ingots with a guaranteed minimum of 99.5% aluminum.

Magnesium. This light metal has come into considerable prominence in recent years, especially in the transportation industry. It is much like aluminum, for which it is frequently substituted. Like aluminum it will burn, particularly in powder form, because it has great affinity for oxygen; it is often used in flares. Magnesium alloys as well as the metal are finding more and more uses in the manufacture of jet engines and other aircraft parts where lightness and strength are critical.

Magnesia, the oxide of magnesium, is a chemically inert white powder that will remain stable at high temperatures. It is frequently used as a refractory; magnesia brick is well known around blast furnaces and steel mills.

The native metal, magnesium, is obtained almost exclusively from sea and saline lake waters. A cubic mile of average sea water is estimated to contain six million tons of magnesium, more than has been produced in the entire history of the world—magnesium is in no danger of exhaustion! In addition to this vast potential supply, another source of magnesium exists in the magnesium carbonate rocks, dolomite and magnesite, which are commonly mined in many places and used mainly as fluxes and refractories.

Magnesium, like aluminum, requires large amounts of energy for its manufacture and can be produced only in areas where low cost energy is available.

Although the metal and its salts are extensively used in local industries, they are of small value in international commerce, not only because of their wide distribution in sea water but also because we still lack knowledge of magnesium metallurgy and do not know how the materials can be cheaply fabricated.

The price of magnesium, which is sold in 10,000-pound lots, has remained relatively stable.

Beryllium, lithium, and titanium. These light metals have been attracting a great deal of attention in recent years. They all require large amounts of cheap power for their production in metallic form and their metallurgy is not yet widely understood; in addition, beryllium and lithium are in short supply. It is anticipated that, as the metallurgy becomes better understood and if supplies become available, these metals will be increasingly useful.

Beryllium is lighter than aluminum and has a high melting point (1278°C). It is, however, brittle and difficult to handle, and some of its salts are toxic to humans.

The principal use for beryllium at present is in the hardening of copper, which it does more effectively than any other substance; in this alloy it is finding many applications. Its salts are also used to coat the inside of fluorescent light tubes, where they serve as an excellent means of dispersing the light. Because of beryllium's toxicity, however, it would be harmful to anyone cut by a broken fluorescent light tube.

The ore of beryllium in the past has been the mineral beryl, a beryllium aluminum silicate that has been recovered in small amounts in Argentina, Brazil, parts of central and southern Africa, and several other places. Beryl is well known in the form of several gemstones, including emeralds, aquamarines, and golden beryls. The supply picture for beryllium has been considerably changed by recent discoveries of low-grade beryllium ores in the minerals phenacite and bertrandite, which are white, either in powder or small crystals, and difficult to recognize. So far these low-grade deposits have been found in Alaska, Nevada, and Utah; additional discoveries will probably be made as beryllium becomes better known.

Beryllium is sold either in the beryllium-copper alloy or as the metal itself, in the latter case in the form of powder, beads, lumps, billets, or rods.

The demand for lithium also has greatly expanded. This metal is recovered from a few lake brines, as at Searles Lake, California, and from complex silicate minerals. Lithium is used in the manufacture of special types of glass and in the electronics and aerospace industries. It is finding further uses as a catalyst in the manufacture of organic compounds, and in alloys. The metal melts at 179°C, but with aluminum forms an alloy that retains its strength up to 400°C, considerably higher than other aluminum alloys.

Titanium, frequently included as a light and therefore nonferrous metal, is discussed in Chapter 4, since it is also considered a ferro-alloy metal.

Environmental problems

In studying the nonferrous metals we must take into account the problem of their smelters. Sulfurous gases composed of sulfur and particulate matter (small particles) blow out of the smelter stacks. The sulfur and dust particles have a deleterious effect upon the area surrounding a smelter. Thus smelters, particularly those that process copper and lead and zinc, are being subjected to strict anti-pollution laws. Many of these laws are long overdue, but their too rapid imposition has resulted in costly difficulties that could have been avoided. Several smelters have been closed; production has been curtailed in others while their owners consider closure; serious economic effects on mine operators and on workers have resulted.

In 1969 only about 4,600 short tons of copper concentrates were exported from the United States for smelting. In 1970 export tonnage jumped to 69,300, of which about 45,300 tons were shipped to Japan. The trend toward foreign smelting seems to be growing. It means that the United States is becoming increasingly dependent upon foreign smelters for vital copper and other metals, a situation which constitutes an unnecessary drain upon foreign exchange and can lead to difficulties of supply (42). It means also the loss of jobs in the United States.

A sign of hope is provided by recent experiments with hydrometallurgical processes of obtaining copper from its ores. They appear to be yielding favorable results and, if they do prove to be successful, will probably displace the present smelting processes because they will cause far less pollution.

The future of the nonferrous metals

Each of the nonferrous metals has its peculiar properties and few generalizations may be made about the group. Indeed, the fact that each mineral is an individual and differs from other minerals has not only contributed to the complexity of the extractive industries and their economics but has also in large part made possible the variety of modern civilization.

As we look at the nonferrous metals, we note that, in spite of the value of mercury and the great demand for aluminum, copper remains the giant of the group. There has been, and still is, an unbelievable amount of searching for copper, and there are nations whose entire economies depend upon this metal. We note also that, although available quantities of most of the nonferrous metals have increased in the past few years, greater availability has increased their usefulness. For that reason the demand has increased, so the metals remain in short supply. Substitutes for all uses of these metals have not been found; there is nothing at present that can for all needs take the place of copper or of mercury.

We note also in how many cases the manufacture of these metals requires large amounts of inexpensive energy.

As with the ferro-alloy metals, the uses of the nonferrous metals will increase with better understanding of their metallurgy—if the supplies increase and if sufficient cheap energy is available. Thus while we may well be disturbed by the prospects of inadequate amounts of copper or zinc, we may anticipate with keen interest the possible future uses of all of these metals.

In recent years deposits of the nonferrous minerals have been subjected to a rash of expropriations. The governments of Chile, Zambia, and Zaire have taken over copper mines; the government of Guyana has nationalized its aluminum ore deposits; other countries are edging close to expropriation.

If the present trend toward nationalization continues, the industrialized nations will have to depend more and more upon purchases in the open market; they will no longer be able to own their mines and gear production to their needs. Serious problems of supply and pricing may be ahead for the nonferrous metals.

7

The Industrial Minerals

The group known as the industrial minerals is large and diverse. Pages could be written to list these minerals and their properties, to describe their distributions and their many uses, and few generalizations apply to all of them. They are usually defined to include a few metals, such as titanium and lithium, but on the whole the industrial minerals are nonmetallic. Sometimes they are called the nonmetallic minerals.

For convenience, the industrial minerals are here divided into two classes, those that enter international commerce in a major way and those that, although they may be exported in minor amounts, are primarily of domestic value. The first category comprises fertilizers, sulfur, asbestos, fluorspar, gemstones, and a few other commodities traded in smaller volume, such as crystalline graphite, the borates, and block mica. The second category contains salt; building materials such as sand and gravel, cement, light-weight aggregates, clays, and gypsum; the abrasives, including garnet, corundum, and emery; absorbents; and many other materials.

Among nonrenewable resources, the industrial minerals have a far greater monetary value than the metals. Even so, and even though it is probably true that any mineral is valuable if it is pure and available in large quantities, more than a few investors have learned to their sorrow that the marketing of the industrial minerals can be very difficult.

Substitutes exist for most of them.

Materials of international significance

Asbestos. Because of its unique fibrous character, asbestos is in demand all over the world. Many varieties are mined. The most useful is chrysotile, which has especially flexible fibers and excellent fire-resistant properties. Long-fiber asbestos is used in weaving fireproof cloth; wearing an asbestos suit, a fireman is equipped to fight a blaze that would sear him if he wore any other material. Asbestos is also used as a heat insulator around pipe, and short-fiber asbestos is utilized for shingles and siding.

The fibrous property of asbestos can be harmful as well as advantageous. In factories during the manufacture of asbestos products its fibers can get into human lungs, constituting a serious health hazard.

This mineral is shipped everywhere from two principal sources of supply, Canada and Russia, with minor amounts coming from the United States, South Africa, and Australia.

There is no good substitute for asbestos, especially the long-fiber chrysotile.

Fertilizers. In a large part of the world, including all the industrialized nations, people are aware of the need for fertilizers to raise good crops. Yet in India and other parts of Asia, and in parts of South America and Africa, where there are hungry people, pathetically inadequate harvests reveal the poverty that prevents them from putting fertilizers on poor soil. And there are still a few places where people are too ignorant to understand the value of fertilizers; on the high Bolivian altiplano bits of colored yarn flutter in the fields of Indians who know nothing about spreading fertilizer but believe that the yarn will ward off evil spirits and bring bountiful crops.

In addition to the hope of education for the ignorant, there is a new hope for improvement in the lot of poverty-stricken people. During the past few years greatly superior short-stem varieties of grain have produced spectacular increases in yield, even from impoverished soils, and appear capable of effectively sidetracking the specter of hunger for some years to come, provided that population expansion is controlled and provided that there is adequate fertilizer, which they require.

At present, in spite of considerable over-capacity in fertilizer-processing plants, a need exists for larger amounts of fertilizers than are being used. Consumption increases slowly; within a few years' time it will again equal or surpass production. If progress is made against the poverty and ignorance that prevent their use in some parts of the world, the demand for fertilizers will increase still further.

Most balanced commercial fertilizers include three principal ingredients, nitrogen, phosphorus, and potassium compounds.

Nitrogen, in the mineral soda niter ($NaNO_3$), was a monopoly of Chile for years, until during World War I German chemists discovered a method of obtaining nitrogen from the air—the nitrogen fixation method. Although small

Fig. 7-1. Potash (K_2O equivalent) production in 1970. "Others" include Spain, 3.2%; Israel, 2.8%; Congo (Brazzaville), 1.3%; Italy, 1.3%. Total production was 20,443,000 short tons. Source: United States Bureau of Mines.

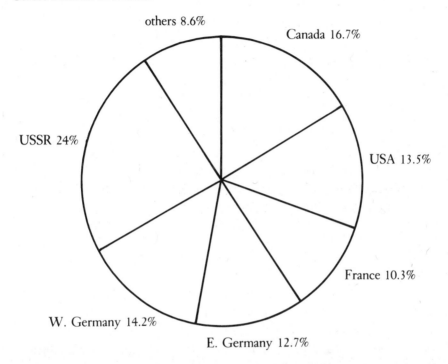

amounts are still obtained from the guano deposits along the coast of Peru and from Chile and other sources, the great bulk of nitrogen used in commercial fertilizers now comes directly from air. The availability of nitrogen is no longer a serious problem, so long as the energy to extract it is available. Here at last is a mineral commodity whose source of supply is not limited to a few places.

Phosphorus as an element, used on the heads of matches, would burn a growing plant just as it makes a match burn; phosphorus in nature is usually combined with oxygen to form a phosphate mineral. Phosphate rock is the source material for most phosphorus and also for the common constituent of fertilizer, which is marketed as superphosphate. From the commercial fertilizers plants can obtain phosphorus slowly enough so that they are not burned.

Extensive deposits of phosphate rock are exploited in the United States, Morocco, Tunisia, Algeria, Senegal, and a few additional nations. Smaller deposits and low-grade phosphatic materials are being mined in other countries, for example Russia, Mexico, Israel, and Jordan. Recent explorations in Australia indicate large reserves in northwest Queensland and Northern Territory.

Phosphate rock is available in quantities adequate for fertilizer needs for many years, but, as with so many mineral resources, areas of deficiency exist. Many of the most densely peopled areas of the world, such as India, China, Indonesia, and Japan, have little or none at all. Japan long ago learned the value of fertilizers, importing large amounts annually, and in the past obtained sizeable quantities of phosphatic materials from several islands in the South Pacific. India, actively encouraged by government and by considerable foreign investments in factories, is only now beginning to expand its consumption of fertilizers. Numerous nations of Africa and South America lack sufficient supplies of phosphate rock to fulfill their needs and can increase their food crops only by importing either phosphate rock or fertilizers.

Phosphorus is also used in the chemical industry and was a common ingredient of the detergents that became so popular because of their effectiveness, greater than that of soap, in removing dirt and grease. Recently, on account of its use in detergents, phosphorus has been the subject of much discussion; in waste waters it becomes a fertilizer and thereby encourages the growth of algae and other unwanted plants, thus choking streams and waterways. The environmental effect of the substitutes used instead of phosphorus in detergents is as yet unknown. They may be as harmful as the phosphorus itself.

Potassium compounds, consisting of potassium and other ingredients such as chlorine, are added to most prepared fertilizers, and between 90% and 95% of the potassium produced in the world goes into the manufacture of fertilizers. The remainder is used for the manufacture of soap and numerous chemicals, for clays and ceramics, for textiles and dyes, and, in the form of potassium nitrate, for explosives, gunpowder, and fireworks.

The value of potassium as a fertilizer was discovered when settlers who cleared their lands and burned off their timber boiled the remaining ashes in large kettles to make a corrosive white solid alkali which they called *potash* because it came from ashes in pots. Potash is potassium carbonate. It is not used in commercial fertilizers, but even now wood ashes are sometimes scattered over the ground to enrich it.

The element potassium is recovered from the brines of lakes such as the Great Salt Lake in Utah where it has been concentrated by evaporation, potassium being the last saline to crystallize when evaporation takes place; or it is mined from salt beds that have been precipitated from the ocean. One of the problems associated with its mining is the disposal of the ordinary salt that must be removed with it.

Potassium is found in the southwestern United States, both East and West Germany, France, Russia, and Poland. Shortly after World War II very large

Fig. 7-2. Fluorspar production in 1970. "Others" include Canada, 3%; Czechoslovakia, 2%; East Germany, 2%; Mongolia, 2%; West Germany, 2%; South Korea, 1%; miscellaneous sources, 3%. Total production was 4,600,327 short tons. Source: United States Bureau of Mines.

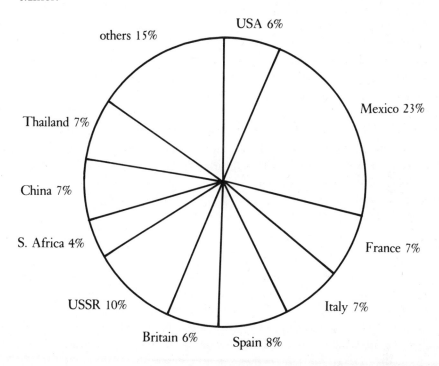

deposits were discovered in Saskatchewan, Canada, and many mining companies hurried into the area in order to obtain property and develop them. This rush into production resulted in a great over-abundance of potassium with a corresponding sharp decrease in price. The glut of potassium which flooded the market caused financial difficulties for many producers, and not until the government of Saskatchewan established a reasonable floor price and restricted the amounts to be exported did recovery begin to take hold. After several years of controlled production the market is now beginning to strengthen. Clearly the over-production of the past few years is temporary. Within a short period of time demands will overtake productive capacity.

Fluorspar. Fluorspar or fluorite, a fluoride of calcium (CaF_2), is an indispensable mineral that most people do not even know. Fluorspar is the principal source of the element fluorine, which is familiar to many people as the material added to drinking water and toothpaste in order to harden teeth. Fluorine is becoming more necessary each year. It has a number of essential uses and large quantities are consumed by the chemical and plastics industries, in the manufacture of aluminum, and in the smelting of steel. With the expansion of the basic oxygen process of steel making, demands for fluorspar, used as a flux, have so greatly increased that they are difficult to meet. Much exploration is being carried on in the search for new supplies.

At present most fluorspar comes from Mexico, Great Britain, Spain, the United States, Thailand, and South Africa, and minor amounts are provided by numerous other areas. The total fluorspar reserves are small. More of this mineral would be used were it readily available.

Sulfur. The degree of any country's industrialization can be judged by the amount of sulfur and sulfuric acid consumed by its industries; sulfur provides an even more exact measure of industrialization than do iron and steel. Sulfur and sulfuric acid are needed by so many industries that without them a nation could hardly be part of the modern world. The element sulfur is used in insecticides, fertilizers, chemicals, explosives, dye and coal tar products, paint and varnish, in the manufacture of pulp and paper (without its bleaching properties, paper would not be white), and in the processing and preserving of foods. Sulfuric acid is used in fertilizers, pigments, rayon and other cellulose products, industrial explosives, textiles, the refining of petroleum, the chemical industry, iron and steel smelting, and other types of metallurgy.

Sulfur is present on the tops of some salt domes. Salt domes result from the compression of salt beds by overlying heavy rocks. The rocks compress the salt until it is forced from the subterranean depths along zones of weakness through overlying rock strata toward the surface, which it may or may not reach.

Salt domes are important, not because of their salt, most of which is not used, but because of two other commodities. At places they contain pools of oil or natural gas along their sides. Since salt is impermeable to fluid under pressure, the salt dome is a trap for oil, which concentrates along its borders. Salt domes are also a source of sulfur. A leakage of gas tends to accumulate near their tops, depositing sulfur in the cap rocks.

The mining of sulfur in salt domes is carried on, not in the usual way with a shaft and underground workings, but by the Frasch process, which utilizes a pipe within a pipe. In the outer pipe, super-heated steam is forced down until it penetrates the rock and melts the sulfur, driving it back up the inner pipe to the surface, where it is collected to solidify in big vats.

Salt domes are not the only source of sulfur; large amounts are obtained as byproducts from "sour" gas—that is, natural gas that contains hydrogen sulfide—and petroleum. More than two million tons of byproduct sulfur come

Fig. 7-3. Elemental sulfur production in 1970. "Others" include Japan, 1.5%, and miscellaneous sources, 1.5%. Total production was 21,748,000 long tons. Source: United States Bureau of Mines.

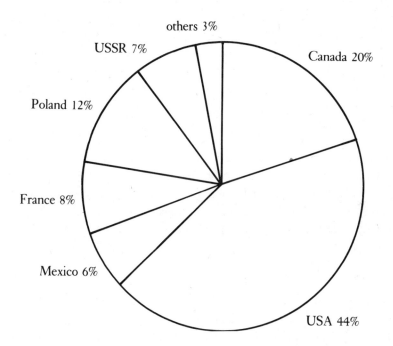

annually from the Canadian gas industry alone. Sulfur must be removed from gasoline and lubricating oil; otherwise corrosive sulfuric acid would ruin the machinery in which they are used. Very large amounts of sulfur and sulfuric acid are also recovered from residual fuel oils, from the desulfurization of coal plants, and from the stacks of smelters in response to demands for cleaner air. The amount of sulfur and sulfuric acid being recovered is so large that there is difficulty in selling them, prices having become too depressed to bear the costs of transportation. It appears that sulfur and sulfuric acid will be in surplus for some years to come, especially as they are being recovered in increasingly large quantities.

Sulfur is also obtained from regions of recent volcanic activity. Italy, Japan, and the high Andean mountains of Chile, Peru, and Bolivia produce small amounts from fumaroles, vents in the earth from which gases issue.

Calcium sulfate minerals, gypsum and anhydrite, have been proposed as sources of sulfur since they can be used for the extraction of this element. However, the process is expensive and the present surplus pushes far into the future any conceivable use of these minerals to provide sulfur.

All the heavily industrialized nations of the world, with the possible exceptions of Russia and the United States, must import at least some sulfur or sulfuric acid. For years the United States exerted the controlling influence on the sulfur markets of the world because the salt domes along the coasts of Louisiana and Texas, some of which are capped with native sulfur, constituted almost the total world supply. Similar salt domes in the Isthmus of Tehuántepec in Mexico have also contributed large amounts of sulfur to world trade.

On account of its widespread use, sulfur is one of the mineral commodities most commonly traded internationally. It is so necessary to industry that when it is in short supply the resulting slow-downs are serious impediments to world commerce.

Gemstones. If the reader were asked to name a gemstone, the chances are that, in the United States and many other countries, he would think first of the diamond. Why? Diamonds may be a girl's best friend just because their value has been so thoroughly and steadily emphasized and maintained, but are they intrinsically superior to other stones? Is their sparkle more attractive than the warm red glow of a ruby, or the incomparable green of an emerald, or the deep blue blaze of a sapphire? It is the eye of the beholder that determines beauty, but certainly, in today's world, the individual's outlook has been strongly affected by the advertising to which he is so constantly exposed. Diamonds furnish a splendid example of long sustained and effective publicity, in this case carried on by the DeBeers Syndicate, which for years has maintained a monopoly of the mining, faceting, and marketing of diamonds.

Esthetic or psychological value has been a fact since some prehistoric man first picked up a pebble because it was pretty. The value of gemstones exists only in the mind; their use as jewels is purely decorative. Yet the value is not only real but great; the gemstones, both precious and semiprecious, are commodities of considerable international significance. In total tonnage traded the amount is small, but in money the value is very large indeed. Several countries depend upon the sale of gemstones for sizeable parts of their international exchange.

The principal exception to the limitation on the uses of gemstones is the important use of diamonds in industry. The diamond is the hardest substance known. It is therefore an ideal abrasive material that will cut every other substance. The stones that are not clear enough to be gems, as well as chips and diamond dust, are used for cutting implements, in abrasives, and as polishing materials. For some of these uses no good substitute exists; the diamond is essential to modern industry. For instance, gem-quality diamonds are needed for wire drawing discs, in the production of wires of standard size. However, synthetic diamonds have appeared on the market in recent years and serve well for abrasives and some industrial uses.

Lately synthetic stones have been introduced for almost all varieties of gems and it is generally recognized that they are superior to natural gems for certain purposes, such as bearings, because they can be made without flaws. Most watches today have synthetic sapphire or ruby bearings. The sale of synthetic stones has not hurt the authentic gem market but, in fact, appears to have improved it, by acquainting people with little known but attractive gems. Here again, as in any consideration of gemstones, human psychology plays a part. A gem can be too uncommon, and it then lacks value; for instance, beautiful rich golden beryl is so rare that few people can own it and not many ever see it. On the other hand, if a gem were to become common, it would lose value.

Nature established the value of gemstones by tossing them out in very small handfuls in very few places. Each gem is found in extremely limited locations, and the locations are different for different gems. A few diamonds come from India, but more are obtained in central and southern Africa and to a lesser extent in Brazil and Venezuela. Recent discoveries of diamonds have been reported in north-central Siberia west of Yakutsk and assure Russia of an ample supply of industrial stones. It is of interest to note that Russia at first sought to market diamonds in the western world by underselling the DeBeers Syndicate. The Russians concluded, however, that they were making no noticeable economic or political gain while they were using up an irreplaceable resource; Russia now markets diamonds in the western world through the DeBeers Syndicate.

Other fine gemstones are of great value in the national economies of the few countries fortunate enough to possess them. The sale of cultured pearls in Japan contributes appreciably to the Japanese balance of trade. Australia has obtained considerable revenue from the sale of its opals, and Brazil has profited by selling aquamarines and other semiprecious stones.

Emeralds and rubies are found in small numbers in several countries. At present the best emeralds probably come from Rhodesia and Colombia and the most beautiful rubies from Burma and Ceylon.

Many other fine stones—sapphire, topaz, spinel, olivine (peridot), golden beryl and other beryls, alexandrite, tourmaline, and several varieties of quartz, to mention only a few—are eagerly sought in their few locales and find ready markets.

Especially in Oriental countries a considerable trade is sustained in jade, much of which comes originally from Burma. Although jade is found in many colors, it is the soft dark green jade that is judged by many peoples in the Orient to be one of the most precious of stones; supernatural powers are frequently attributed to it. Certainly some of the intricately carved jade objects are true masterpieces of the sculpturing art and are justly prized by their owners.

The United States produces very few high-quality gems but is by far the largest consumer in the world.

The gemstones are mined, except for pearls, which are found in oysters. It is fascinating for anyone who has only an esthetic appreciation of these materials to think of them as minerals, coming from underground mines or being washed out of stream gravels. Only a variation in internal atomic structure makes a diamond different from two other materials to which it is chemically similar—lamp black and graphite.

The wheels of industry would turn without gemstones, with the possible exception of industrial diamonds, and no person thinking of the relative value of minerals would consider them, as a group, essential. They cross our pages evoking a flash of loveliness, trailing those intangible qualities described as "esthetic and psychological value" and "beauty." The real and definite worth of these elusive qualities is shown by the fact that gemstones are, like the quite different minerals we are discussing in this chapter, of significance in international trade. Asbestos, fluorspar, and sulfur, fertilizers and jewels, they are all minerals wanted by people everywhere in the world.

Others. Many other industrial minerals enter international trade widely but in smaller amounts than the better known commodities such as gemstones and fertilizers. The fact that the quantities traded are small does not necessarily mean that the materials are unessential. Many have special uses for which they are pre-eminently suitable.

Among the products included here are the micas—the isinglass of industry— particularly the block micas used in the manufacture of bridges or supports in radio tubes, as the dialectric in capacitors, and as insulation in many pieces of electrical equipment; the borates, widely used in the chemical industry; and flake or crystalline graphite, used in foundry facings, crucibles, lubricants, batteries, "lead" pencils, steel making, and for many other purposes.

Materials of domestic value

Salt. Since salt is such an important mineral commodity, it is ironic that the disposal of superfluous salt from the processing of potassium compounds and the salty edgewater of many oil fields poses problems for the potassium and petroleum industries. But salt is a cheap commodity. The salt from such sources (although sufficient to be a potential cause of serious pollution) is generally too far from a market to be of value because transportation and recovery costs would be too high. At many oil fields the salty water is forced back underground, exerting pressure on the gas and oil and moving them toward the wells where they can be recovered.

Salt, or sodium chloride, is obtained by the evaporation of sea water. It is also obtained from the mining of the mineral halite in quarries and underground mines, as in Michigan. Salt is usually found in subterranean beds, once the floors of lakes or oceans, from which water has evaporated. The expression "back to the salt mines" to describe an unwilling return to work comes from the days when slaves were sent into mines and probably thought salt mines the most trying of all.

Salt is found in the same geographic locations as potassium (central Europe, southwestern United States, and Saskatchewan, Canada). Salt can also be recovered from sea water by evaporation, although many countries, like Japan, have so much rain that they find it advantageous to import salt from arid coastal regions, for example the western coast of Mexico, by cheap ocean freight. The discovery and development of more and more salt beds and salt domes continue.

Salt is required by all vertebrate animals; all people, from the most primitive to the most sophisticated, eat it, and salt in blocks is commonly supplied for cattle and other livestock. In addition, salt in large quantities is used in the chemical industry and now has a winter use in northern latitudes, where it is spread on snow and ice to make them melt. This practice has been criticized because the salt damages vegetation and contaminates water supplies.

Salt has many industrial uses in which it is consumed in immense quantities. Over two thirds of the domestic salt output of the United States, which totals thirteen million tons a year, go into the chemical industries for numerous items, including chlorine (necessary for the manufacture of many chemicals), sodium (used as a water softener), and caustic soda (necessary for refrigeration). Salt is needed also for soap, including detergents; for food processing; for the preservation of food, including fish; for ice manufacturing; and for ceramics, including glass.

The importance of this mineral goes far back in human history. The early Germans fought for their salt streams, attaching a religious significance to them. Salt was used as an offering by the ancient Hebrews, Greeks, and Romans, and the term "salt of the earth" first referred to Christ's disciples. To "sit above the salt" signified in medieval times the possession of a place of distinction, while those of less importance in the eyes of the host were placed "below the salt."

This mineral was important in world and domestic trade in ancient times, when salt and incense were the chief commodities and influenced early trade routes. Salt formed the basis of trade between ports in Syria and the Persian Gulf; salt was traded between Aegean ports and the coasts of southern Russia; the old salt mines of northern India were a trade center before the time of Alexander the Great; and the Via Salaria (the "Salt Road"), along which salt was carried from Ostia to the Sabines, is one of the oldest roads in Italy.

Salt was often a state monopoly and hence a source of revenue to kings and governments in ancient and medieval times; it remains so in Italy to this day. Cakes of salt were, long ago, used as money in Abyssinia and Tibet, and even now, in some primitive areas of central Africa, the finest gift is a bag of salt. In civilized countries today, almost everyone, by using such expressions as "the man is worth his salt," demonstrates the long-lasting value of this industrial mineral.

Building materials. Building materials include a wide range of ordinary commodities such as sand and gravel, cement and light-weight aggregates, clays for the manufacture of brick and tile, and gypsum for the manufacture of wallboard. Most of the building materials must be obtained near the places where they are used; they are all low in price per unit and cannot stand the costs of long-distance transportation. Gypsum and clay are cheap because they are abundant; when, as is seldom necessary, they must be transported far, their price rises. If gypsum and clay were less common, their price would be higher. The rarer china clay, in white porcelain, costs more than clay for common brick.

With the advent of extensive suburban developments and the imposition of zoning laws, ready availability of sand and gravel has decreased greatly. People do not want to live near gravel pits, and the result is that the sources of these materials are being pushed farther and farther from city centers and thus from their places of use. This means added building costs. In some areas, shortages are expected to become increasingly critical. Many builders are turning to the use of crushed stone to replace the needed sands and gravels. In places where there is no sand or gravel, such as parts of eastern Brazil, crushed stone provides serviceable roads but its cost is higher in both money and human energy.

Most people accept sand and gravel with little thought as to their availability or value. Tremendous amounts are used; in the United States alone in 1969 some 937 million short tons were used at a total value of one thousand seventy million dollars. In these days of expanding population, a good sand or gravel quarry is an extremely valuable property.

Cement is widely manufactured and finds new and growing uses in construction and paving. Cement plants are located near limestone deposits and their outputs are sent to communities that are not far away, since long-distance transport is seldom profitable. In the cement plant, limestone mixed with clay is burned in kilns and carbon dioxide gas driven off; large amounts of lime dust are generally poured into the atmosphere along with this innocuous gas. Environmentalists have severely criticized the cement industry, and anyone who has lived near a cement plant can testify to the lasting corrosive effects of the dust on paint and other materials. Newer cement plants control most of the dust and do a much better job of keeping surrounding areas attractive.

Sand, gravel, and crushed rock are mixed in large amounts with cement to form concrete, but in recent years considerable quantities of light-weight aggregates have been substituted for these materials, especially in the construction of high-rise buildings. Materials used for light-weight aggregates include pumice, perlite, and vermiculite. Perlite is a form of volcanic glass that contains small amounts of water. When heated the water expands and the rock "pops" or breaks into a porous light-weight material much like pumice. Vermiculite also contains water; physically it resembles mica, and when heated it expands slowly. Perlite and vermiculite are used not only in concrete aggregates but also in plaster aggregates, accoustical materials, and agriculture. Such materials cannot bring high prices. Their mining, treatment, and transportation costs must be kept at a minimum.

Gypsum, so necessary for wallboard, is for the most part obtained from massive beds of solid rock that are widely distributed. Gypsum is mined in at least twenty states of this country.

Clays are of many kinds and have varied uses. In the building industry those used in the manufacture of bricks and tile are among the most useful. Modern brick and tile plants require large amounts of easily mined clays as well as supplies of low-cost energy for burning the clay. Many specialty glazed bricks and tiles require unusual processes and materials.

Refractories. Refractories are able to retain their shapes and chemical properties when subjected to high temperatures. They are particularly useful for lining furnaces and for other processes that require inert materials resistant to heat. Most of us are familiar with the fire-resistant or refractory bricks that line fireplaces.

The refractories include a wide variety of materials whose use depends upon the place and type of process. Magnesite and certain varieties of clay are used to manufacture many firebricks. There are also high alumina and silica (acid) refractories and basic (non-acidic) refractories. Other minerals and compounds have been used, generally for special purposes, in the refractory fields.

Others. Many other industrial minerals are of domestic significance. They include a wide range of abrasives such as corundum, emery, and garnet, absorbents such as diatomite, the bleaching clays used for adsorbing coloring matter in oil, and numerous materials that defy classification. The latter include many ceramic clays, bentonite (used as a binder in the manufacture of iron ore pellets), kyanite, andalusite, other oxides of aluminum (used in the manufacture of spark plugs and insulators), feldspars (used in ceramics), and pure, colorless quartz sands suitable for the manufacture of glass.

8

The Petroleum Crunch

~~~~~~

If you were in the United States late in 1973 when the Arab countries imposed sanctions on the sale of petroleum, what was your reaction? Whom did you blame? The Arabs? The oil companies? The government? Yourself? (53)

There are many sides to that question. Some have been over-emphasized; some have been ignored. We plan to show some aspects that have not been widely taken into account because some facts have not been widely known. Thus you may be better informed and have a more balanced view than that of the general public when, cold, partially immobilized, and forced to pay more at the service station, it cast its blame on those it considered responsible for causing a shortage of gasoline, heating oil, and electricity.

The Arabs

Restricting the sale of their petroleum appeared at first glance to be a political act by the Arab nations, to enlist allies in their effort to regain land lost to Israel in the Six Day War of 1967. When the United States was not supported (except by The Netherlands) in seeking a détente between the Arab countries and Israel, the Arabs resumed petroleum sales to all NATO countries except The Netherlands and the United States. Actually any political gain was an extra dividend to the Arabs, secondary to the major benefit for which they had long been planning.

For the Arabian action was primarily economic. Arabs already had large corporate holdings, in some cases controlling interests, in multinational oil companies operated by the British, Dutch, and Americans. Most large oil companies are multinational. For instance, Aramco's four American oil companies are controlled by stockholders from many nations, including

Fig. 8-1. The world's demand for petroleum in millions of barrels per day. The demand projected for the future cannot of course be expected to develop as shown unless petroleum resources provide adequate reserves. Source: Marshall Ayres.

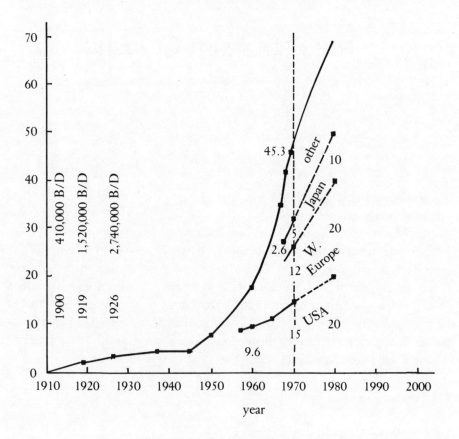

Arabian countries. Owning the oil as well as having a powerful corporate voice in the companies producing it, the Arabs decided to get more revenue from it. When political sanctions were eased, those companies that bought petroleum had to pay to the nations owning the oil taxes on the posted price* materially higher than before. The price paid to owners also jumped. Both the price and the taxes the producers must pay are reflected in the cost to the consuming nations. Thus those countries that could buy petroleum had to pay considerably more than they had in the past.

* The posted price is that figure established by an oil-owning nation as the basis for taxes paid locally by the companies producing the oil. The highest posted price early in 1974 was Libya's, $18 a barrel. From country to country, the posted price varies, and so does the local tax rate based on it. Whatever the posted price and tax rate may be locally, this is only the first cost.

The oil companies

Most people could understand that a nation acts politically and economically for its self-interest, even if the action turns out to be short-sighted. When it came to blaming the oil companies, however, the public was both uninformed and inconsistent. These were the general charges:

• The oil companies acted in collusion.
• They were making excessive profits at the expense of the small independent distributor and the consuming public.
• They had spent their capital in recent years in foreign countries, thus contributing to an adverse balance of trade.
• They were not prepared to meet a shortage of imported oil.

Collusion. The oil industry is similar to any other large and complex industry. It is made up of people and run by people. Of the twenty-five largest companies in the United States, some are public spirited, others will seek all possible profit.

In one respect, these companies collaborate: they maintain a lobby before Congress, as do other industries and various groups. But this lobby was unable to prevent the reduction of the depletion allowance for exploration and development, or the imposition of quotas on the amount of oil that might be produced domestically or imported.

Those who believed that the oil companies were locked in a tight conspiracy did not know a number of facts. The explorations and geologic maps of individual companies have been treated more confidentially than State secrets. Bidding for development rights has been highly competitive. In Alaska in 1968, for example, the bidders lost millions of dollars in interest because all bidders were required to deposit 20% cash in Alaskan banks before all bids were opened simultaneously. Had there been collusion, there would have been no need for losing bids and lost interest. Had there been collusion, the expensive advertising and the gasoline price wars of the past would also have been unnecessary. Had there been collusion, how could the largest company have gained less than 9% of the total market?

The oil business has been like professional football or baseball. Before the public, the companies have acted together, as a league, in order to gain public support. The individual companies, however, have been like individual teams. They may make trades, or deal with each other, but on the field they have been ferociously competitive.

Excessive profits. Petroleum has been as large a business as almost all others combined. Profits from the petroleum business should be large simply because of its size. Yet during the decade of the '60's, according to the index of the Department of Commerce, the oil industry in the United States earned about 10% on capital invested, while other industries earned more than 11%. Almost half of the profits after taxes of oil companies have been distributed in dividends—on which income taxes have been paid. The amount of federal income taxes paid by these corporations has about equaled net profits. When local and payroll taxes were added, the amount of taxes exceeded net profits. Excise taxes collected on sales were additional payments. How could our government have operated without that tax income?

In 1973 the profits of the oil companies without doubt increased greatly in percentage over 1972. But the profits in 1972, and indeed from 1969 to 1973, were low, probably running just over 1¢ a gallon. It is sometimes misleading to see the figures of an oil company's gross profit or its amount of capital. In such a competitive business, large amounts of capital are necessary, to bid on development rights in Alaska ($1 billion) and the Gulf Coast ($1.48 billion), for instance, or to gamble on essential but dubiously successful exploration and drilling. One has to remember that estimates generally run to a billion dollars a year spent on the drilling of dry holes, and that a refinery costs several hundred million. To attract risk capital, a corporation must pay good dividends and show promise of making good profits. The alternative, in the petroleum business, is too little capital to explore or develop.

When we in the United States were getting gasoline for about 34¢ a gallon, direct taxes, federal and local, took a major share of that amount. Another cost was retail service at the station, whether independent or company-owned. Any "excess" profits came out of the 15¢ or so per gallon that remained as gross income to the oil companies. How much would you call excess? When you consider that the corporate net income was 10% on capital invested, you can realize that the maximum possible reduction would have been less than 1½¢ a gallon if all profits had been wiped out. During recent years, how much difference would a reduction of a cent or two in the price of a gallon of gasoline have made to our pocketbooks?

Capital invested abroad. It is true that oil companies have invested large capital sums in foreign countries, more than for exploration and facilities at home. They have done so in order to increase their profits, the purpose of any company. If a few companies have at times appeared arrogant, the record indicates that the industry as a whole has not been avaricious. If an oil company did not take advantage of opportunities abroad, its competitors would, and the administrators who were remiss would be liable to stockholder charges of incompetence. Any successful company must put its capital where the returns are best.

Fig. 8-2. The world of known oil reserves. Sources: Exxon Corporation and the *Oil and Gas Journal.* Copyrighted and used with permission.

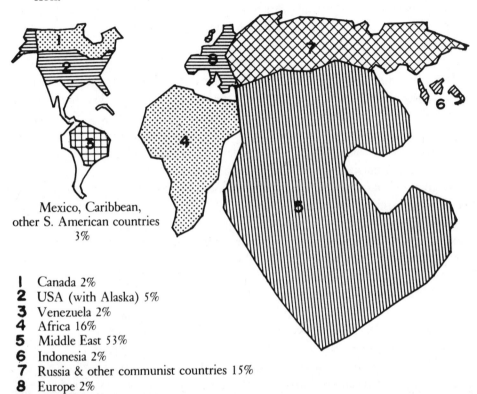

Mexico, Caribbean,
other S. American countries
3%

1 Canada 2%
2 USA (with Alaska) 5%
3 Venezuela 2%
4 Africa 16%
5 Middle East 53%
6 Indonesia 2%
7 Russia & other communist countries 15%
8 Europe 2%

Why should the returns be best abroad? Why have our oil companies gone elsewhere to explore, drill, and produce?

They were forbidden to explore or to develop what are known to be the most promising areas in the United States:

- The oil shales of the west.
- The continental shelf off our north Atlantic shores.
- The west coast of Florida.
- Most of the coast of California.
- Much of the northern slope of Alaska.

Areas available for exploration for oil were sharply restricted in recent years as environmentalists prevented offshore drilling and as more and more public lands were declared off limits—in Alaska, for example, where a billion dollars had been spent for development rights and there will have been neither return of that capital nor any profit for at least eight years. Where is the search for oil being welcomed today in the United States?

And we must not forget that capital investments abroad have brought home dividends and profits on which taxes have been paid.

Lack of foresight. At various times spokesmen and publications of the oil companies have tried to call attention to the problems of future supplies. Yet restrictions have continued on development and on exploration.

Until recently, the government enforced import quotas on oil; it did not relax them until the supply situation became critical. No oil company could afford to expand a refinery or to build a new one unless it was assured of an adequate supply of crude oil to permit operation of the new facility—and quotas on imports had limited the supply of crude so that there was no need for refineries. When oil shortages were at last recognized officially, the refinery capacity could not be expanded as much or as quickly as necessary.

Even now, the construction of new refineries is being strongly opposed in areas where they are most needed.

Another restriction, the prevention of offshore drilling along the entire eastern coastline, resulted in rather bitter feelings in the southern oil-producing states, where people asked why they should be forced to send fuel oil to New England when they themselves needed it. If New England needed fuel oil—which it did—its people should have been willing to look in their own back yard rather than to depend on other parts of the nation, where people likewise do not particularly welcome offshore drilling and refineries.

The oil companies have been restrained in many other ways, such as in the construction of pipelines and access roads and the provision of adequate port facilities. Modern tankers of 400,000 tons and more are becoming well established in the industry. The United States does not have a single port capable of accommodating these giants.

How could United States oil companies be prepared to meet a shortage of imported oil? They were hampered by restraints on domestic exploration and development. Their imports had been restricted by quotas. They did not have freedom to move. They could not explore because areas at home were closed to them. The situation was an example of an old adage: If you don't look for oil, you're not going to find it.

In addition, the oil companies have now lost much of their control over sources of supplies, not only in the Arabian nations but in Venezuela, Canada, and other countries where the companies invested capital and skill in exploration and development. The industry is tightly regulated. No longer can a company increase or decrease the supply of crude oil without permission from the country that contains the oil field. No longer do decisions on where to ship and what price to charge remain freely in the hands of the officers of an oil company.

The price jump. An additional criticism was leveled at the oil companies early in 1974. Why did prices jump, people asked, when according to figures released by the oil companies themselves stocks of crude oil at hand were at least as large as they had been a year earlier? "I have to pay 60¢ a gallon now when I buy gasoline," said the average citizen angrily. "But it's just for the profit of the oil companies. There isn't any shortage."

Many reasons caused the price jump. The Arabs had increased prices by a factor of 3 to 5—and some of the Arab countries were shipping oil to the United States in spite of the embargo. Because Venezuela, Iran, Nigeria, and Indonesia had followed the Arab lead in raising prices, the oil companies faced higher prices on all imported oil.

In addition, the decrease in supplies of natural gas had put an extra demand on petroleum suppliers from public utilities, which needed more natural gas as fuel.

Still another contributing factor was the forecast of a bitter winter, which had caused our northeastern states to insist that more petroleum be distilled for fuel oil—and when the winter turned out to be relatively mild the oil companies' storage tanks were left brimming with higher-quality fuel oil instead of gasoline.

In 1973 the government, fearing political repercussions, removed its quotas, so more of the expensive oil could enter the country.

Earlier in 1973 the government had also removed price controls on expanded domestic production. That had made possible the reopening of thousands of small, old wells producing a few barrels of oil a day, wells that could not be operated economically at the prior gasoline prices. These wells could be reopened and operated only at higher costs.

Thus the stocks of crude oil that were at hand had cost the oil companies more than the stocks of the previous year. The oil companies were forced to pass the higher cost along to the consumer if they were to stay in business. The stocks were available, but only at a higher price, and their availability did not mean that the nation was not facing a shortage. With the probability of even more price increases on the part of foreign oil-producing nations, with the extra costs required to obtain oil from domestic sources that were bound to become less and less economic to operate, who could tell how high the price might eventually be for a gallon of gasoline? And it was certain that shortages of natural gas and fuel oil could lead to a reduction in gasoline supplies.

The public criticism most dangerous to the public itself was the belief that in fact no shortage existed. This belief was based on lack of knowledge.

Public relations. Whose fault was it that the public was unknowledgeable? Would the oil companies have had to shoulder the blame for the petroleum

Fig. 8-3. Per capita use of energy in the United States from 1955 projected to 1985 in terms of barrels of petroleum or their equivalent in British Thermal Units. Source: The Chase Manhattan Bank.

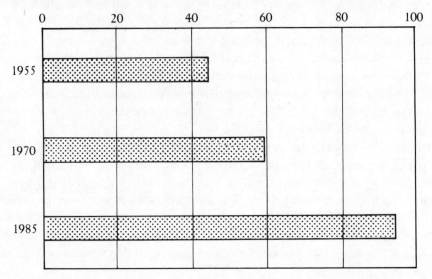

crunch if over the years they had been as skillful in relations with their public as they had been in technology?

A good example of the ineptness of the oil companies came early in 1974 when, under pressures from the public to explain the petroleum shortage, many of the companies responded with full-page ads in newspapers across the country. So complex was the content of most of these explanations, and so formidable the size, that probably not one newspaper reader in a thousand read them with understanding. Earlier the companies had given the public very little information; now they gave too much too late.

Perhaps the nature of the oil business explains much of the failure of oil companies to maintain good public relations.

An oil company represents the magnification of all the problems of extractive industries. For example, exploration and development require large capital risks based on educated guesses. The business constantly demands fierce competition—between company and nature, between company and competitors, or between company and the government of the land where oil is found. Although oil executives have been negligent in this respect, it is easy to see how their daily concerns left little time for educating the public to the special problems of the petroleum industry.

Fig. 8-4. Domestic production and consumption in quadrillions of British Thermal Units of energy in the United States from 1950 through 1971. The lower line shows production. The upper line shows consumption. The divergence of the two lines shows the amount consumed in excess of the amount produced domestically. Source: F. L. Hartley in *Seventy Six*, Union Oil Company.

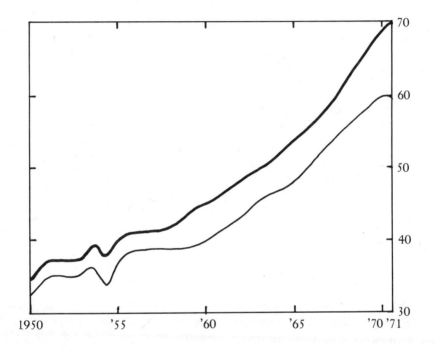

In addition, in its retailing aspect the oil industry has never come close to the public, dealing only indirectly, through service station operators. It has become apparent that those with the technological skill necessary to operate an oil company were not used to applying skill in retail activities. Oil company executives were used to battling, and on a large scale. They ignored or mishandled the average citizen. For that reason, the average citizen knew nothing about oil, either its problems or its significance. When Mr. Ralph Nader said before Congress on January 15, 1974, "The world is literally drowning in oil. Geologists say we have used less than 25% of the world's oil reserves. The reserve figure supplied by oil companies actually represents only about 10 percent of real proven reserves*," the press and the public accepted a lawyer's statements. Who the geologists were who provided him such a figure has not been ascertained.

* Mr. Nader should have said "resources" (Chapter 1).

The oil industry's lack of foresight and imperviousness to public opinion were to blame for the public's acceptance of this and similar misstatements. Yet in the energy crisis that has still to develop, we will badly need the industry's technological skills, unmatched by any foreign competitor. We may blame the oil industry for its failure to carry on consistently an intelligent and forthright campaign to educate the public; certainly it could hardly have built up a worse image or poorer credibility if it had deliberately set out to do so. But if we blame the oil industry, we should not kill the goose, foolish only because of astigmatism, that has been laying golden eggs for some time now. It has provided us with good gasoline for less than half of what consumers paid in Japan and Europe.

The American consumer's dollar. Compare the figures of sales at a retail gasoline station for 1968 and for 1973. 1968 (when there were neither shortages nor complaints about prices) was the most profitable year for the oil companies until 1973. Estimating that ten gallons cost at the gas station $3.40 in 1968 and $5.10 by the beginning of 1974, we asked, "What happened to those sums?" Two major oil companies—Shell and Standard of California—answered by giving us their figures for 1968 and 1973.

Shell Oil, always near the top among retail sellers of gasoline in this nation, is about 69% owned by Royal Dutch, a European company which is not subject to our laws and taxes. Standard of California is completely American in ownership, although many of its shareholders live abroad. Its operations are international; for example, it is one of the four American partners of Aramco.

The figures of these two companies, like those of any American corporation, are subject to review by the Internal Revenue Service, the Securities and Exchange Commission, and independent financial analysts such as Standard and Poor's, and both companies are subject to United States corporate laws.

We asked each company for figures, for each of the two years, that could be summarized in a simplified form:

(1) What were gross revenues (principally sales of products)?

(2) What were costs of operations and of products (other than taxes and new capital investments)?

(3) What taxes (income, excise, and all others) were paid?

(4) What profits were paid out to shareholders as dividends?

(5) What profits were held by the corporation?

(6) How much capital was invested (in refineries, for example)?

We decided to combine the figures supplied by the two companies in order to present a better cross-section view of the industry. We then computed items (2), (3), (4), (5), and (6) as percentages of item (1). The results are shown in Fig. 8-5(A). Next we applied these percentages to prices paid for ten gallons at the service station in 1968 and 1973. The results are shown in Fig. 8-5(B). Gross revenues combined were $8,414 million for 1968. By 1973 they were 74.4% greater. The allocations of items (2) through (6) are in terms of 1968's gross revenues at 100%, and are so applied in Table II. This does not affect the percentage allocations of items (2) through (6).

• The price of gasoline at the service station did not increase at as great a rate as did costs to the companies.

• The increase in costs over the five years was due to many factors, such as inflation, higher taxes, increased demand, and higher costs of crude, especially from abroad. Therefore prices and gross revenues increased. The State of California received a royalty of $9 a barrel, which was 22¢ a gallon, for crude pumped from its tidelands, until federal price controls reduced the amount. The state then resorted to court action in an attempt to re-establish the higher rate.

• All sales were assumed to be gasoline, none the cheaper residual products. Any resultant distortion of figures favors the gasoline purchaser, not the companies.

• No allowance was made for the margin of an independent dealer. (One independent dealer in California says he counts on 5¢ a gallon profit after his costs.) The country has approximately 200,000 service stations, of which 95% have been operated by independent businessmen.

• Since 1960 the debts of the oil industry as a whole relative to its capital have risen from about 10% to over 33%. Normally when any corporation expands expensive facilities, as in building a refinery, it will borrow money or issue stock or combine the two methods of raising funds if they are not available from surplus. Because since 1960 the oil companies have made smaller profits on their capital investments than have other industries, new stock has not generally been an attractive buy to investors. Therefore, the oil companies have had to borrow what they have not had in surplus capital (item (4) in the tables)—and interest rates have been high.

• Dividends paid have been less than those of other industries, even lower than interest paid by savings and loan associations. Standard and Poor's quotes this "yield" at 12/31/73 as 3.5% for Shell and 4.9% for Standard of California. The varying amounts of taxes paid by individuals on these dividends are not ascertainable.

• Royal Dutch Shell, a foreign corporation, made more profit in 1973 than did its domestic affiliate Shell Oil. Royal Dutch's profit increased 153%, from 5% of revenues in 1972 to 9.4%. Shell Oil's profit increased 28%, but relative to its revenues only from 5.4% to 5.8%. Similarly Aramco, in which the Arabs are the largest stockholders and Standard of California a minor one, and which operates abroad, made a much greater 1973 profit than Standard's. Is this

Fig. 8-5. An example of the oil industry's operations (**A**) for its two most profitable years and (**B**) in terms of purchases of ten gallons of gasoline at the retail pump. The differences between the figures for the two years are (1) increase in revenues of 74.44% and (2) that percentage allocated 55.83 to costs, 13.70 to taxes, 0.65 to stockholders, 4.26 to the companies' undistributed profits. These allocations have been applied in part (**B**) of the table. Figures are based upon the gross revenues of two major companies in 1968 as 100%. Sources: Shell Oil Company, Standard Oil Company of California, and a United States citizen.

A. *Two oil companies' figures combined*

	1968	1973
(*1*) Gross revenues (chiefly sale of products)	**100.00%**	174.44%
(*The revenues were divided as follows*)		
(*2*) Costs of products and operations	68.74%	71.41% . .
(*3*) Taxes paid (income, excise, etc.)	22.18%	20.57% . .
(*4*) Profits distributed to shareholders	4.39%	2.88% . .
(*5*) Profits added to companies' funds	4.69%	5.14% . .
Total	100.00%	100.00%
(*6*) New capital expenditures relative to (*1*)—(compare with (*5*) above) . .	(12.42)	(7.19) . .

B. *Purchase of 10 gallons, 1968 and 1/1/74*

	1968	1973
(*1*) Cost at the retail pump	$3.40	$5.10
(*Divided as in* **A** *above*)		
(*2*) Costs of products and operations .	$2.34	$3.64 . .
(margin for independent dealer omitted)		
(*3*) Taxes paid by the oil companies .	.75	1.05 . .
(*4*) Dividends paid to shareholders .	.15	.15 . .
(*5*) Profits added to companies' funds	.16	.26 . .
Total	$3.40	$5.10
(*6*) New capital expenditures (compare with (*1*) and (*5*) above) . .	(42¢)	(43¢) . .

because foreign consumers are willing to pay higher prices than we for products from crude? Were we getting a bargain at the pumps early in 1974, after our long waits in line?

If we blame the oil companies to the extent of driving them out of the country, or out of business, are we not cutting off our own noses? Can a

Fig. 8-6. When, early in 1974, the profits of domestic oil companies were announced, the increases were in the U.S. Congress termed "obscene," an opinion widely quoted in the news media. The increases in 1973 over the prior four years were large only in terms of percentages.

This table includes audited figures for the two companies of *Fig. 8-5* plus figures for Exxon and Gulf: the gross revenues of 1969 are used as the base of 100%. The table, in the form of *Fig. 8-5*, permits some interesting comparisons: Taxes paid *vs.* dividends and retained profits (taxes paid by recipients of dividends are not included, of course); new investments in facilities such as refineries *vs.* retained profits (this explains the companies' increasing debts); retail price *vs.* what it would have been if all retailed profits had been eliminated; profits from 1969 through 1972 *vs.* those of either 1968 or 1973. These comparisons and other factors contradict statements that 1973 profits were anything but reasonably better in that they allowed a safer operating margin.

A. *Four oil companies*	*1969*	*1972*	*1973*
(1) Revenues	100.0%	130.9%	160.9%
(disbursed as below)			
(2) Costs	62.6%	61.9%	60.2%
(3) Taxes	29.9%	31.5%	31.4%
(4) Dividends	4.7%	3.8%	3.2%
(5) Profits held	2.8%	2.9%	5.2%
Total	100.0%	100.0%	100.0%
(6) Into new facilities		(9.6%)	(8.5%)
B. *Ten gallons retail*	*1969*	*1972*	*1973*
(1) We paid	$3.40	$3.40	$5.10
(disbursed as below)			
(2) Costs	$2.13	$2.10	$3.08
(3) Taxes	$1.02	$1.07	$1.60
(4) Dividends	$ .16	$ .13	$ .16
(5) Profits	$ .09	$ .10	$ .26
Total	$3.40	$3.40	$5.10
(6) New facilities		(32¢)	(43¢)

government bureaucracy, with the "advantage" of no competition, do well with an operation that demands complex technological ability? Has Congressman Claghorn yet read the Paley report?

Fig. 8-7. Economic and political crises during the coming decade will center around the uneven distribution of resources over the earth, the rising costs of inadequate supplies, and the profits of the multinational companies that supply the demands. The case cited below prepares one to listen and read critically.

The largest corporate enterprise in the world outside the U.S.A. is known as the Royal Dutch Shell "Group." Operated from London, it is controlled by the Dutch and British. Some major entities in the Group are Royal Dutch Petroleum, Shell Trading and Transport, Shell Caribbean Petroleum. These and others in the Group are independent of U.S.A. control. The Group has a 69.4% interest in Shell Oil Company, a U.S. corporation subject to U.S. laws and taxes. In this simplified analysis one sees the worldwide nature of the extractive industries, and the fact that our domestic prices (and the profits from them) are not the highest in the world. The Group obviously explores, produces, transports, and can be expected to sell where prospects are best.

A. *Shell Oil Co. (U.S. corporation)*	*1972*	*1973*	*Increase*
Gross revenues (millions) . . .	$ 4,850 . .	$ 5,750 . .	18.6%
Net profits	$ 260 . .	$ 333 . .	28.1%
Profits relative to revenues . .	(5.4%)	(5.8%)	

B. *Royal Dutch "Group" (international)*	*1972*	*1973*	*Increase*
Gross revenues (millions) . . .	$14,400 . .	$19,000 . .	31.9%
Net profits			
from 69.4% of Shell (U.S.) . .	$ 180 . .	$ 231 . .	28.3%
from other sources	$ 524 . .	$ 1,549 . .	195.6%
Total	$ 704 . .	$ 1,780 . .	152.8%
Profits relative to revenues . .	(4.9%)	(9.4%)	

Government

In June of 1952 the Paley report, prepared by the best scientists in the country, was presented to President Truman and made available to all members of Congress and governmental agencies who should have been concerned. This report, in five volumes, documented the many problems of raw materials. One volume dealt specifically with the problems of energy supplies as of 1952 and in the future. Yet the Paley report's facts were ignored by government. Nothing was done to increase the supply of energy or to establish an energy policy.

LETTER OF TRANSMITTAL

—————

THE WHITE HOUSE,
Washington, July 1, 1952.

Hon. SAM RAYBURN,
Speaker of the House of Representatives,
Washington, D. C.

MY DEAR MR. SPEAKER: I am transmitting to the Congress the report of the President's Materials Policy Commission, "Resources for Freedom." Our knowledge and understanding of the materials position of the United States and of its allies throughout the free world will be considerably increased by the detailed review which has been prepared by the Commission. This is a document which deserves the most careful study by every member of the Congress, and I hope each one of them will take the time to familiarize himself with its contents.

This report, the fruit of months of intensive study by an independent citizen's group aided by experts drawn from Government, industry, and universities, shows that in the past decade the United States has changed from a net exporter to a net importer of materials, and projects an increasing dependence on imports for the future. The report indicates that our altered materials situation does not call for alarm but does call for adjustments in public policy and private activity.

In more than seventy specific recommendations, the Commission points out the actions which, in its judgment, will best assure the mounting supplies of materials and energy which our economic progress and security will require in the next quarter century.

I am requesting the various Government agencies to make a detailed study of these recommendations, and I am directing the National Security Resources Board to assume the responsibility of coordinating the findings and of maintaining a continuing review of materials policies and programs as a guide to public policy and private endeavor. As the need arises for legislation to solve materials problems affecting this Nation and other free nations, appropriate recommendations will be made to the Congress.

It is my hope that this report and the actions which may be taken as a result of it will contribute significantly to the improvement of this Nation's materials position and to the strengthening of the free world's economic security, both of which are the continuing objectives of United States policy.

Sincerely yours,

HARRY S. TRUMAN.

(From the volume on energy of the Paley report, when President Truman sent the volume to Congress)

The executive branch. Should we blame the executive branch of government? During the twenty years that government has been informed of an impending oil shortage, the United States has had five different presidents. Often Congress was not politically in sympathy with the president and was unwilling to enact laws or allocate appropriations for his programs. A need calling for a long-range policy cannot be answered by a changing branch of government that has not consistently had the power to command the laws and appropriations such a policy requires.

Congress. The legislative branch of government has had over the past decades the continuity of responsibility that the executive branch has lacked. Some senior senators and representatives have twenty or thirty years' experience, holding office through four or more presidential terms, and now head committees that determine important actions. Congress is charged with passing laws, making appropriations. Presumably it does so after careful study; at least that would be indicated by the number of congressional hearings. It has at its command all the national funds, and all the national brains, those of specialists whom it may call upon to study problems and recommend plans.

This powerful body is composed almost entirely of people trained in law. Even a congressman's assistants and staff members are usually lawyers. That means that the members of Congress are trained in the art of compromise, and are most impressed by those gifted in persuasion. It also means that they, regardless of their legal competence, are in general ignorant and uncomprehending about any question relating to the earth's resources, including petroleum.

To whom has Congress listened, *with comprehension,* regarding any problem of natural resources? It did not listen to the specialists who wrote the Paley report. It did not listen to experts from the National Academy of Sciences who were knowledgeable about the situation. It did not listen to economic geologists who forecast the need for planning. It did not listen to the petroleum suppliers, the oil companies. If it did hear any warning given by anyone who knew anything about the earth's resources, it took no action.

Congress has listened principally to three groups of those trained in the art of persuasion, three kinds of opinion-formers: growth economists, television commentators, and extreme environmentalists.

Growth economists. These men have for years spoken of an ever expanding economy. Their thesis has been that a growing population means increased demands for products. Increasing demands mean increased production, and therefore increased payrolls and national income. All that increase means in turn more sales of products and more corporate income and federal taxes. Such a glowing picture of a future without economic recessions proved a welcome sight to a Congress whose competence was legal and political.

The ones who tried to point out the flaw in this picture were those who knew something about the earth. They knew that the earth is limited in what it can yield in space, materials, and the energy to find, extract, process, and distribute material products. They knew that the limits of the earth make an ever expanding economy impossible. But they were not the ones to whom Congress listened.

Television commentators. There used to be several hundred newspapers giving a variety of opinions and full coverage of news. Now we have only three major television networks with selected topics covered in short daily summaries. Thus television commentators have an impact on opinion that did not exist a decade and more ago.

The commentators and their production editors do a good job, as they must in order to obtain the ratings that determine advertising income. They are suave, persuasive, and in some cases most intelligent, as well informed as possible considering their deadlines. They pass their convictions on to the public and to Congress.

These opinion-formers, in general, accepted the growth economists' beguiling surety about the ever expanding economy of the future. They also accepted the wishful thinking of environmental extremists.

Environmental extremists. Like all opinion-formers, environmental extremists are persuasive. Frequently they too, like members of Congress, are lawyers, and similarly ignorant of the basic facts of natural resources.

They are idealists. They believe in the preservation of all land and natural objects. They do not acknowledge that change is a law of life and of the environment.

Popular and influential, the environmental extremists persuaded Congress, as we have noted, not to permit development of the great reserves of petroleum that exist in the United States, from Alaska to Florida and from the Atlantic Coast to the Pacific.

Who was to blame?

It appears evident that, although the executive branch of government, the oil companies, and the Arab countries that shut off supplies all shared in the blame for the petroleum crunch, the burden of blame should fall most heavily upon Congress. The United States never has had, and does not now have, a long-term, carefully considered energy policy. Exploration and development are still handicapped. Supply is proving to be an even greater problem than the environment. We need much more than a system of temporary allocations and poorly conceived crash programs originating in many bureaus. We need action by Congress that will permit and implement coordinated energy planning by means of an established, consistent, and recognized policy.

But does the blame rest there? Who chooses the members of Congress, who votes? Who, finally, is to blame for what Congress does or does not do?

Significance of the petroleum crunch

International significance. We have limited our discussion to the United States, but the petroleum crunch affected every country in the world. It was felt most strongly by the countries whose industries depend in large part on imported petroleum for energy. Western Europe, for instance, relies on coal and oil for its energy, yet has no proven oil reserves of any extent. About one third of the industrial energy used by England is imported petroleum. Over 99% of the petroleum used by Japan is imported.

When non-Arabian oil producers followed the Arabs in raising the price of their petroleum, energy costs rose noticeably in England, western Europe, and Japan. The average share in the Japanese stock market dropped from about 5,500 to 4,500 yen in one day. Italy was faced with a deficit of many billions of lire. England shortened its work week, partly due to a strike by coal miners but partly because of the high cost of petroleum. Not only industries based directly on petroleum but industries based on petrochemicals, such as drugs, synthetic fabrics, chemicals, and plastics, had to meet increased costs for their raw material.

The intense and growing competition for oil was shown clearly by the actions of Japan, which had obtained about 80% of its requirements from the Persian Gulf states. In order to continue to obtain supplies, Japan entered into contracts that resulted in an increase in the prevailing price of crude oil. This was the situation in Kuwait, for example, where Japan obtained offshore concessions. Japan also entered into purchase contracts with the government of Abu Dhabi at prices above prevailing rates. Any increase in the price of crude oil in one country of the Middle East generally spreads quickly to all producers.

Economists familiar with the petroleum industry estimate that by 1980 something like 45 to 50% of the oil consumed in the United States will be imported—unless, of course, other sources of energy are developed. By the year 2000, an annual deficit in our balance of trade is predicted in the order of $60 billion, and by far the largest item in this deficit would be petroleum. No economy, no matter how strong, can long withstand deficits of this size.

Even under present conditions, when imports of oil in the United States are still relatively small, large amounts of so-called Eurodollars have accumulated abroad, especially in Europe and Japan. These dollars result from a continuing deficit in balance of trade and payments. In 1972, the dollar deficit was $6.5 billion, and imported fuels accounted for 60% of that amount. The situation became so critical for some nations that they were reluctant to accept more dollars. This is understandable; what could they do with a large accumulation of steadily depreciating dollars? Suppliers of petroleum preferred to sell to other nations with sounder currencies. Another and somewhat beneficial result of the surplus of dollars was that foreign companies and governments began to return money here by investing in our refineries and industries. The oil-rich nations were particularly interested in United States refineries and marketing; Saudi Arabia, before the imposition of sanctions, had made a tentative offer to guarantee delivery of our petroleum needs in return for a substantial participation in our refinery business.

One of the saddest results of higher oil prices is that the poorer nations suffer most. In India, government officials fear that the country's $71 billion five-year development plan may be doomed before it can begin. South Koreans worry that their country's industry in synthetic textiles is threatened by the growing shortage of imported petrochemicals. Kenya faces a loss in its principal industry, tourism, because the rise of oil prices in Europe has curtailed travel.

Seventy to eighty countries must import petroleum. They paid $2 billion in 1970; the same amount of oil would have cost $10 billion at the prices prevailing in January of 1974—and what was to prevent the Middle East oil-producing nations from raising prices further?

The economies of the world's less developed countries in Asia, Africa, and Latin America are already suffering from the rising price of oil. Unlike wealthier countries, they lack the financial reserves to pay for increases, and they cannot decrease any inessential uses of oil because they are using it only for essential industries and agriculture.

Another question concerns the eventual safety of the Arab oil-producing nations, which own over 50% of the proven reserves of petroleum but constitute only 2% of the world's population.

Significance for the future. The ultimate effects of the petroleum crunch cannot now be foretold. Its greatest importance was that it apparently warned people and made them at last realize that supplies are limited. At least it gave them a chance to reach that realization. For that reason, people in the United States, and in other countries where suddenly there was not enough petroleum at a "reasonable" price, owe the Arabs some thanks (though not too many). They let us see what could happen. They gave us a glimpse of future shortages and a perhaps final chance to awaken from our lethargy before it is too late.

The petroleum crunch was no reason for either acute pessimism or fitful optimism. It called for a realistic appraisal of the facts.

The petroleum crunch was not the energy crisis. It foreshadowed the energy crisis.

How probable is it that there will be a real energy crisis? How soon might it occur? What can we do about it?

To find any answers, one needs to see the whole picture of which petroleum is only a part. One needs to know about energy.

Postscript

As this book goes to press, the draft of an agreement has been accepted by oil-importing nations which consume more than 75% of the petroleum used. They are the Common Market group of western Europe (except France), Canada, and the United States. Japan was not a participant, probably because of its extreme dependence on imported oil.

The immediate purpose of the agreement is clear enough: to counter the actions of the oil-exporting nations in reducing production and raising prices. The long-range effects on the oil exporters may be indeterminable for some time.

Two less obvious aspects of the agreement are especially significant. This is the first occasion on which so many powerful nations have agreed to yield to an international body some of their sovereignty; under the agreement, a representative governing board may make decisions binding on all participants. Further, the agreement makes evident that, to combat the universal struggle against inflation, the first requirement is availability of inexpensive energy.

9

Energy I

~~~~~~

Energy is the *capacity to do work*. All human beings possess it, although they may spend it in playing instead of working, and they must spend it in sleeping, thinking, breathing, merely existing. Actually manual energy is energy in its most inefficient form because, in comparison with any other form, it takes more time and effort to produce less results. You need no explanation of that statement if you have ever had to climb stairs in a high building when an elevator was not functioning, or if you have mowed a lawn first by hand and then with an electric lawnmower, or pushed a handcart after driving a truck. Every housewife knows that washing machines, dishwashers, and vacuum cleaners do a more thorough and healthful job and save hours of drudgery. Electrical appliances are superior in efficiency. For instance, each electric blanket means that two or three extra blankets need not be manufactured, and food can be preserved in refrigerators and freezers to an extent that was undreamed of before their invention. The undeveloped countries that must still depend upon manual energy have a higher cost for many commodities and a lower standard of living than the industrialized countries that are able to utilize machines.

In electricity energy is measured in *watts*. As you probably know, electric circuits accommodate a certain number of volts, and the brighter light uses more energy, or watts. A volt is the force (electromotive force) required to drive a current of one ampere through a resistance (in the electrical circuit) of one ohm. The work thus performed is measured in watts. Therefore, a watt is the power of one ampere flowing across a potential difference or resistance of one volt in the circuit.

The energy obtained from the combustion of gasoline, fuel oil, diesel oil, and other petroleum derivatives, as well as coal and natural gas, is measured in units of heat. Since heat is one form of energy, the work done or the energy released by the burning of a fuel can conveniently be expressed in heat units. The common units are the British Thermal Unit, or *BTU*, and where the metric system is employed, the *calorie*. A BTU is the amount of heat required to raise the temperature of one pound of water one degree fahrenheit. A calorie is the amount of heat required to raise the temperature of one gram of water one degree centigrade. One BTU equals 252 calories. A barrel of crude oil contains on the average about 6,000 BTU's. During the burning of any fuel, complete combustion and use of the potential energy is never attained; some part is always dispersed into the surroundings and lost to man.

Horsepower is the measure of the work a machine, for instance an automobile, may perform. One horsepower is the equivalent of 550 foot-pounds of work per second, which means the energy required to raise 550 pounds one foot in one second.

Electricity

Electricity may be called a secondary form of energy. It is generated by steam from the burning of a fuel or by falling water in a hydroelectric plant at a dam. While the total amount of energy used in the United States has increased at the rate of about 4% annually, the amount of electricity used has increased at more than twice that rate, doubling every eight to ten years.

In recent years, shortages of electric power, or "brownouts," have become more and more common in the United States. These annoying shortages have been caused as much by the lack of sufficient electrical power-producing capacity to meet the demands of the country as by the lack of primary energy such as coal or fuel oil or hydroelectricity.

Falling water

Hydroelectricity is among the cleanest and most desirable sources of energy used by man. Power from falling water leaves no residue, no scars on the land other than the required dams, and emits no gases into the air.

In some parts of the world, such as Norway, Japan, and western Canada, where water supplies are plentiful and stream gradients steep enough to permit the development of a high hydraulic head, large proportions of the energy consumed come from falling water. In the United States about 5% of the energy used in 1970 was from hydroelectricity.

However, hydroelectric power is not as cheap as it may appear to be. In the United States, we have made its cost seem lower in many instances by allocating a large part of the expense of dam construction to navigation, flood control, and recreation.

Moreover, power from this source cannot be increased appreciably in the industrialized parts of the world, where its utilization is now almost maximum. Not many streams in the United States offer possibilities for further power development, and conservation groups strongly urge the retention of "wild rivers" so that at least a few can be maintained in their original condition.

Only in the tropical rain forests and mountainous areas of some emerging nations is abundant potential water power still available for development, from rivers such as the Amazon, Orinoco, and Magdalena in South America and the Congo and Ivindo in Africa. One cannot stand on the bank of any of these mighty rivers without feeling the tremendous force of its power; the Congo River at the Stanley rapids is an awesome sight even at low water. The harnessing of these streams and use of their energy for the good of mankind could completely change the economies of large parts of the earth. Although the cost of erecting necessary dams and power plants would be great, in many cases far beyond the financial capabilities of the underdeveloped nations where the rivers are situated, so much energy would be obtained with such minimal maintenance expense that the unit cost of the power would probably be low. Many industries that require very large amounts of electrical energy, such as those that manufacture aluminum metal, might be more effectively located near these cheap sources of hydroelectric power, especially since in the same places jobs are needed and large amounts of aluminum ore await development.

Petroleum and natural gas

The organic (or fossil) fuels—petroleum, natural gas, and coal—now furnish by far the major supply of energy used in the world. The total energy used in the United States in recent years is attributable 44% to petroleum and 31% to natural gas.

In terms of collective capital invested, petroleum ranks among the top industries of the world. In international trade, petroleum and its products account for about half the total volume. Petroleum is number three in value of the industries of the United States, following agriculture and all utilities combined. About 90% of the transportation used in the United States is run by petroleum. Possessing about 6% of the proven petroleum resources of the world, the United States uses about 33% of the world's petroleum output. This comes to approximately 5.5 billion barrels of crude oil a year, or about 26.4 barrels per capita of the nation's approximately 208 million people. A barrel of crude oil contains 42 gallons of the hydrocarbons that constitute the petroleum plus various minor but important impurities.

Petroleum has become essential because it has been plentiful, cheap, and easy to transport.

Fig. 9-1. A projection of imports of petroleum products into the United States and the cash outflow for these products. "C" is Canadian crude; "F," foreign offshore crude; "PI," imports of other petroleum products. The projection of cash outflow was made in 1972 and estimated the cost per barrel at $3 to 1975, $3.50 by 1980, and $4 by 1985. Source: Shell Oil Company.

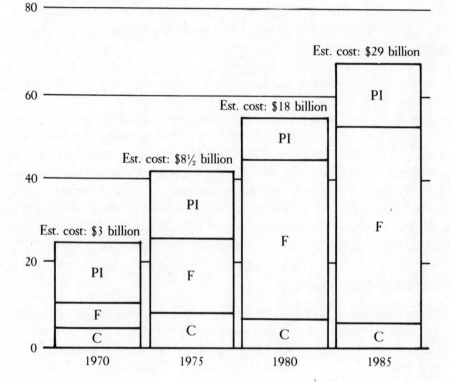

Petroleum has been known to man since ancient times, when it was used to coat walls and sailing vessels and as mortar and a fire weapon. It was also burned in lamps, though not often. In North America the Indians used it to make paint and for medicine, considering it magic; it was the "Genesee oil" and "Seneca oil" that early settlers bought from the Indians to use medicinally.

In the nineteenth century the use of petroleum to burn in lamps was extended in Europe; Prague illuminated its streets by this method. Petroleum was obtained from the tops of streams, from holes in the ground, or even from shale.

In 1859 a man named Drake drilled a producing well in Pennsylvania and started the modern petroleum industry at a place that later became Titusville. Other wells followed in the area, known as Oil Creek; their principal product was kerosene, used in lamps. Ownership was determined according to the "law of capture" by the surface location of the petroleum, without regard to its underground source.

Oil was used chiefly for kerosene until the invention of the gasoline engine.

At present petroleum is more than a source of energy; products made from it have many uses. One of the most valuable is as a lubricant, essential to modern industry. Numerous varieties of lubricants are available on the market, designed for every conceivable condition of operation and kind of weather. Without them most machinery and vehicles would squeak and grind to a halt. For their preparation, large amounts of petroleum are demanded; uranium and other nuclear power sources will not serve this purpose.

Synthetic fibers, plastics, drugs, and various other products are also based on petroleum. In fact, several oil fields have been set aside by their operators as more valuable for petrochemical purposes than for energy. A petrochemical is any product of petroleum except energy—such as a lubricant, a plastic, or a detergent. Judging by the steady stream of new petrochemical products that reach the market and the amount of research conducted by the chemical industry, one would think that it would expand at a rapid rate and in the future demand a greater share of the petroleum output. Many nations are building petrochemical plants; competition for the international market is keen. We must expect to reduce our use of petroleum for energy if in the future the chemical industries require more of it.

There are approximately three thousand items obtained from the refining and treatment of crude oil, and, since residual fuel oil is relatively unprofitable, American refiners have preferred to use their crude oil for gasoline and other highly refined products. Thus oil companies on the east coast of the United States, especially New England, have had to import to obtain the residual fuel oil which constitutes the major fuel used there for electrical generation and in addition provides inexpensive heating in large commercial and institutional structures. The reliance of the eastern seaboard on low-sulfur residual fuel oil has risen spectacularly in the past few years, especially in those locations where it is replacing coal because of regulations that limit the emission of sulfur oxides and particulate matter. About 90% of the residual fuel oil consumed in the United States originates outside of the national boundaries, approximately 80% coming from the Caribbean and roughly 10% from the Eastern Hemisphere. The consumption of residual fuel oil by United States east coast utilities plants increased 63% between 1967 and 1969, and has increased since then, at a somewhat slower rate. Although the growth in demand continues, a lighter distillate oil or coal will have to be substituted as residual fuel oil becomes scarce.

The cleanest fossil fuel is natural gas. In these days of consciousness of the quality of the environment, gas is preferred over other fossil fuels for that reason, and also because it is cheap to transport and distribute.

As increasing amounts of natural gas have been sought by consumers throughout the world, there has been a rapid increase in international shipments of gas in refrigerated tanks. This has resulted in the conservation and use of gas from many fields where it was forced back underground to help in augmenting the flow of oil to the surface, or even burned as a waste material.

Reserves of natural gas have not kept up with the growing demand. In the United States the ratio of proven reserves to production has fallen continuously over the past few decades, although additions to reserves were sufficient to add modestly to the total until 1967. After that year, and for the first time, reserves fell absolutely. Reasons for the shortage have been the subject of considerable debate; regulations formulated by the Federal Power Commission on explorations and on amortization of risked capital are pointed out as the greatest single cause. Regulated ceilings on the price of the commodity are said to be so near exploration, production, and distribution costs that profits are insufficient to permit exploration and development, even though knowledgeable people claim that large supplies of natural gas are yet to be found.

In 1971 the Potential Gas Committee in the United States, which makes a survey of the gas supply situation every two years, stated that plenty of natural gas was waiting to be discovered in this country—offshore, in the continental fields at depths greater than fifteen thousand feet, and in Alaska. But development of wells in all of these places is expensive, and private industrialists say that they cannot undertake such a development program so long as a squeeze exists between cost and price.

Whatever the reason for the shortages, potential customers in some areas are now unable to obtain supply commitments. As reserves of natural gas have dropped, public utilities have increased their demands on fuel oil supplies.

Where petroleum and natural gas are found. In the past the petroleum production of the world has come mainly from four areas—the United States; the Middle East, including Saudi Arabia, Iran, Iraq, and several small but wealthy sheikdoms nearby; Venezuela; and Russia. To these must now be added places where more recent discoveries have been made: Indonesia, Libya, the North Sea, Nigeria, Colombia, and Canada, all of which have fields of economic value. Many smaller fields contribute to overall production and to the industrialization of the nations that control them, including areas in Rumania, China, Sakhalin, Australia, Gabon, Chile, Bolivia, Peru, and Mexico; there are many others. Petroleum is not centered in any one part of the world, although many countries are almost entirely dependent upon imports.

Deficiencies of petroleum and natural gas exist in some of the most heavily

Fig. 9-2. Crude oil production in 1970. "Others" include Canada, 3%; Algeria, 2%; Indonesia, 2%; Nigeria, 2%; Abu Dhabi, 1%; miscellaneous sources, 11%. Total production was 16,689,617,000 barrels. Source: United States Bureau of Mines.

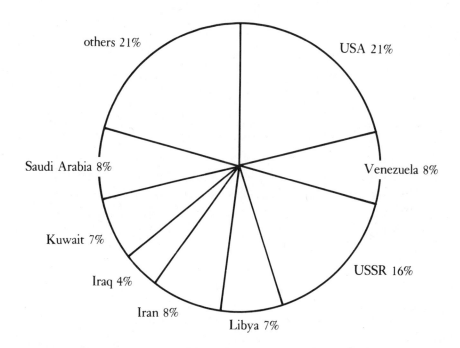

industrialized nations of the world, Japan and most of western Europe; some of the most densely populated nations, including India, Pakistan, and much of southeast Asia, contain almost none (5). These and many other nations, including Brazil, the Philippine Islands, and some in central and southern Africa, require imports. Among the industrialized nations, the United States has been particularly fortunate in having large supplies of petroleum and natural gas available within its borders, and for many decades was a net exporter. This situation no longer exists and the United States now imports more than 35% of what it consumes. Should the oil recently discovered off the east coast of Canada prove to be part of a good field, supply problems along the eastern seaboard might be alleviated for a while.

A large part of the earth's more accessible terrain has been examined and re-examined in the search for petroleum and natural gas. Discoveries in these places have not kept up with demands, so the search is being extended, with considerable success, into the more remote and inhospitable areas of the earth, into oceans and in tundras of the far north. A great deal of offshore exploration is carried on now in many places throughout the world. Usually it is in depths of water less than six hundred feet, the maximum at which offshore wells have proved profitable, but the depth is slowly being extended as technology and equipment improve. The Arctic shoreline area has been another focus of recent exploratory effort.

Exploration in these undeveloped regions has required new techniques of drilling, extraction, and transportation, all of them expensive. For initial development a well in Louisiana may cost about $50,000; a well at the same depth in northern Alaska may cost $3,000,000 or more.

Not only is exploration extending to hostile environments; it is also pushing deeper and deeper into the earth. On February 6, 1972, a well drilled in Oklahoma by the Lone Star Producing Company passed the previous record of 28,500 feet; the target depth of the hole was reported to be 31,000 feet.

Very large reserves of oil shale in the Rocky Mountain states and of tar sands in western Canada should assure North America of future supplies when the cheaper products are exhausted or supplies cut off. Efforts have been made in the past few years to recover oil from the tar sands, although restrictions imposed by the Canadian government have tended to hold back development. The oil shales so far have been generally ignored, partly because of the high cost of recovery and partly because no corporation could afford the expense of operating on public lands on the terms required by the United States government. By early 1974, however, sixteen companies had pooled their research efforts and announced their intention to develop a viable oil recovery process, using shales on private property.

The processing of the tar sands and oil shales requires large amounts of heat and water. The energy of heat is expensive even though the availability of oil from the sands and shales reduces the cost. Many oil shales are situated near the continental divide where water is in short supply. In addition, there is the problem of disposal of waste once oil has been extracted from either sand or shale. The waste fills much more space than the original deposit because the process of extraction has resulted in less compact materials. Possibly subterranean explosions can be used to fracture the shale and release the oil into the cavern left by the explosion. If this technology is developed, the oil will be released from the shale at relatively low cost and the amount of waste material will be minimized.

How oil and gas are obtained. Natural gas is found in the same kind of geologic environment as petroleum. Often they occur together, although one may be found in a field from which the other is absent.

Both oil and natural gas accumulate underground in pools; the fluids are in permeable rocks and are contained or trapped by overlying and surrounding

Fig. 9-3. The sources of energy used in the United States in terms of millions of barrels of crude oil per day. This projection was made in 1972. Source: Shell Oil Company.

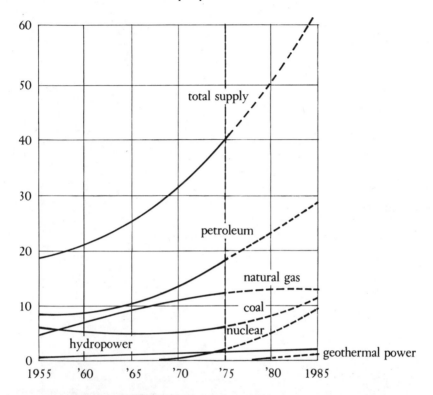

impermeable clays or rocks. Oil or gas fills the spaces between grains of the porous rocks. Actually an oil pool consists of a rock with only the intergranular spaces holding the oil. Most oil and gas pools are underlain by water—the edgewater of the oil fields—which has a tendency, by exerting an upward pressure, to keep the petroleum and gas contained.

Petroleum probably takes less than one million years to form. Once it is formed it must be concentrated in pools if it is to be economically available. The accumulation of petroleum trapped in pools may take 100,000 years or more (*38*, pp. 522 et seq.).

When oil and gas are encountered in a well, they may flow to the surface. Often, however, the oil does not flow and therefore must be pumped in the same manner as water from a well. The oil well is lined with steel casings which are perforated in appropriate places so that oil can flow into the well. As with water wells, techniques are designed to make the material near the hole more permeable in order to permit the fluids to accumulate more readily in the well pipe.

Fig. 9-4. Ownership of rights to the oil and gas fields in the North Sea. Source: *Engineering and Mining Journal*, McGraw-Hill. Copyrighted and used with permission.

Increasing the flow of oil by forcing gas or water undergound is one of the most common methods used in a procedure known as *secondary recovery*, which may be employed in order to conserve and recover as much oil as possible when the output of a well starts to decrease. The gas forced back underground through auxiliary wells maintains a pressure toward the producing well, causing the oil to flow in that direction. Similarly, water flooding underground increases the pressure on the oil and causes it to flow into the wells. The reader will remember from the discussion of salt that the edgewater of oil fields is usually so salty and impure that it cannot be emptied into surface drainage but must be forced back underground, thus aiding in the recovery of more oil. Steam may also be forced underground to aid in secondary recovery. Secondary recovery methods in many of the older fields have resulted in saving large amounts of oil that otherwise would have stayed in the ground. In some instances secondary recovery has exceeded the primary one because of recent improvement of technologies. However, even with the best of secondary recovery methods, large quantities of petroleum remain in the rock when an oil field is abandoned. There are varying estimates of the portion of oil that cannot be recovered by current techniques; most authorities accept from 30% to 50% of the original pool.

Secondary recovery methods are also used to obtain the greatest possible amount of natural gas.

In the United States alone, in 1969, 32,173 wells were drilled for oil and natural gas. Of these, 14,368 produced oil, 4,083 produced gas, and 13,722 were dry holes (*44*). The amount of oil found per foot of drilling has decreased sharply in recent years. In 1973 in the United States, 26,592 wells were drilled. Of these, 9,902 produced oil, 6,385 produced gas, and 10,305 were dry holes. Those producing oil totaled 44,641,949 feet, or an average of 4,510 feet per hole. Those producing gas totaled 35,600,331 feet, or an average of 5,420 feet per hole. The unproductive holes totaled 56,149,172 feet, or an average of 5,441 feet per hole, thus accounting for more than 41% of the total footage drilled. In 1973 about 30% of the total number of wells (8,008) were drilled by wildcatters, those venturesome outfits willing to take risks on long shots.

Oil and natural gas on the international scene. Because of the dominant role petroleum and natural gas play in determining the prosperity or poverty of a country, each country with a potential supply seeks self-sufficiency. Oil and gas have been absolutely essential to survival as a modern nation, to industrialization, and to a high standard of living. Thus most governments closely control all aspects of the petroleum and natural gas industries, and in some countries government monopolies control exploration, production, refining, and marketing.

OPEC. The Organization of Petroleum Exporting Countries comprises eleven nations that produce about 77% of the world's petroleum and control about 93% of the world's petroleum exports: Saudi Arabia, Iran, Iraq, Kuwait, Abu Dhabi, Qatar, Libya, Algeria, Venezuela, Indonesia, and Nigeria. Most of these nations are classed as underdeveloped; they are trying to improve their economic positions, and to do so they need money. What better way to obtain it than from their petroleum?

Early in 1971 the OPEC countries demanded an increase in royalties and taxes from the foreign oil-producing companies that were developing oil fields in OPEC countries. Although the companies agreed to a settlement which meant their payment, by the end of a five-year contract, of an annual amount fifteen billion dollars greater than the amount paid in 1971, the Secretary General of OPEC announced soon after the settlement that member nations also wanted direct ownership participation in oil concessions within their territories.

Two years later the OPEC nations began to implement their decision. Iran was nationalizing its petroleum industry and moving toward entering all phases of production, refining, and marketing. Iran, Saudi Arabia, Kuwait, and several other producing nations had begun to seek long-term, stable contracts which would enable them to expand their ownership and increase their profits gradually and in an orderly and legal manner. On the other hand, Libya and a few others resorted to quick expropriation without compensation, using their abundant resources and favorable economic situation as political clubs to enforce their demands.

As early as 1970 Libya, a major supplier of low-sulfur fuel oil, curtailed its production in order to create shortages in Europe and thus force up the price. In 1969 the posted price at the well-head was $1.007 per barrel. When Libya permitted the resumption of full-scale production in March of 1971, the posted price increase was as much as 90¢ for a barrel of crude oil. By September of 1973 the price was $6 per barrel. By early 1974 it exceeded $10. The posted price is the basis on which taxes or revenues are paid to the owners (Chapter 8).

Also in 1970 further shortages were caused purposely by a "mysterious" cutting, with a bulldozer, of the Trans-Arabian pipeline that crosses Syria and connects the oil fields of the Persian Gulf with the Mediterranean Sea. After nine months of negotiations, the owner, the Trans-Arabian Pipe Line Company, was allowed to repair this vital pipeline, but of course the cost of transmission then increased.

The loss of the pipeline capacity, combined with the closure of the Suez Canal resulting from the Israeli-Arab conflict, necessitated increased tanker shipments of oil around the southern tip of Africa in order to maintain European industry. The Trans-Arabian pipeline is 1,024 miles long and can carry 450,000 barrels of crude oil daily from the Persian Gulf to the western terminus at Sidon, Lebanon. The estimate is that about six times as much tanker capacity is required to ship this amount of Persian Gulf oil to Europe around Africa as to ship it from an eastern Mediterranean port (63).

Increased demands for tankers, caused by the closures of the Suez Canal and the Syrian pipeline and the curtailment of production in Libya, were responsible for a serious shortage of shipping and exerted further upward pressure on the price of petroleum. After the reopening of the pipeline, the resumption of production in Libya, and the OPEC signing in 1971 of the five-year agreement with the oil companies, and with the building of supertankers, the petroleum industry entered a period of uneasy stability. It was apparent, however, that in all OPEC nations a transition from private ownership to state participation if not state ownership was taking place. During the early summer of 1972, Iraq expropriated the oil fields of the north, claiming that production there had not been what it should. Representatives of the oil companies pointed out that production from these fields had been curtailed because production from other areas was cheaper. The issue was resolved when Iraq, retaining the fields, paid compensation for them.

Toward the end of 1971, when Great Britain removed its troops from several of the small island sheikdoms in the Persian Gulf, Iran took possession, much to the displeasure of Iraq and Libya. Libya quickly announced nationalization without compensation of the British oil holdings in Libya; Britain retaliated by court action in an attempt to prevent Libya from selling oil from the fields that were formerly Britain's. In March of 1972, Libya stated that it had made arrangements with Russia for the disposal of Libyan oil. What then? Russia has no need for Libyan oil. The picture is still incomplete and can be clearly seen only against its political background.

The only Western Hemisphere member of OPEC, Venezuela, where concessions had been granted into the 1980's, recently announced nationalization of all oil properties by 1975. About two thirds of the 3.7 million barrels a day output from Venezuela are exported to the United States, primarily in the form of heavy industrial fuel oil, for which Venezuela is the chief United States source of supply. The proposed Venezuelan laws would gradually impose additional government controls over the petroleum industry even before nationalization. Some company officials claim that these would amount to total intervention of the government in every administrative and operational activity. One clause would even extend government ownership to office buildings, employee recreational facilities, and houses that are now owned by the oil companies. At present more than 80% of the profits obtained from Venezuelan oil reverts to the government.

Venezuela also plans to develop projects for the liquefaction and export of natural gas, particularly to the United States. With this stated objective, a law was passed in August of 1971 nationalizing the country's natural gas industry.

The Venezuelan legislation is widely attributed to the current trend toward nationalism, but unfortunately also to anti-North Americanism. Most of the oil and gas concessions are owned and operated by United States companies.

Nearly all of the fields of Venezuela are past the peak of production; reserves are estimated to be sufficient for no more than ten years at the present rate of pumping.

Any stability for the petroleum industry ended in 1973 when hostilities broke out between the Arab nations and Israel. Late in October, the Arabs announced plans to curtail their output of oil 5% each month until Israel withdrew from occupied Arab lands and the Palestine question was settled. At the same time King Faisal of Saudi Arabia announced suspension of all exports to the United States. The other Arab countries followed his lead, and the United States found itself cut off from supplies which were answering about 8% of its needs.

The increase in prices for all petroleum products originating in OPEC nations was felt most immediately and severely in Europe and Japan, the principal consumers of Middle Eastern oil. The United States would ordinarily feel the pressure more slowly and to a lesser extent. However, price boosts by Canada as well as by Venezuela and Indonesia caused prices here to approach the level of those in Europe. The continuing increases in the price of petroleum and its products, and of natural gas, and curtailment of supplies of these commodities, can have profound effects on United States industry, already subjected to a serious cost-price squeeze. Higher prices may stimulate further attempts to develop the Canadian tar sands and cause work to start on the Rocky Mountain oil shales, as well as give an added impetus to the installation of nuclear power plants and the search for other sources of usable energy.

OAPEC. The Organization of Arab Petroleum Exporting Countries is composed of Saudi Arabia, Libya, Abu Dhabi, Algeria, Bahrain, Dubai, Qatar, and Iraq; Tunisia and Oman have applied for admission. OAPEC started in 1968 with the declared purpose of helping the producing Arab nations to diversify their sources of income through participation in the entire spectrum of the world petroleum industry (*54*). In March of 1972 the OAPEC nations agreed to establish an Arab Tanker Company, to be owned jointly by the eight countries, with an initial capitalization of one hundred million dollars, governments of the countries providing 51% of the money and the remainder, 490,000 $100-par shares, being offered to the public (*Wall Street Journal*, 14 March 1972).

The members of OAPEC also plan greater involvement in exploration, production, transportation, refining, and marketing. They seek full development of their oil resources, higher revenues, and closer controls over domestic production. However, until the Arab-Israeli conflict of 1973 is finally resolved, all further developments of Middle Eastern oil will probably be delayed. No matter what the outcome of the conflict, higher prices are in store for consumers.

The trend toward more intensive participation of all governments in all aspects of the petroleum and natural gas industries could well lead to a change in the corporate structure of the large international petroleum companies. Several of these companies are at present as nearly multinational as any that exist, and with greater participation by governments they may become the first truly multinational organizations, with primary ties scattered around the world.

Transportation of oil and natural gas. The business of transporting petroleum, petroleum products, and natural gas is one of staggering proportions, since approximately half the volume of shipping in the world is devoted to these commodities and their derivatives.

The Suez Canal was closed as a result of the Middle Eastern war in June, 1967. Prior to that time a few very large tankers were in use for transporting petroleum on long hauls between deep-water ports. The closing of the Canal necessitated a great expansion in the tanker business, giving impetus to the construction of more large ships. This was encouraged by a record consumption of oil in Europe, where industry was booming. By the middle of 1971, many of these giant tankers were in service, and even larger ones were planned. According to *Atlas* (4), quoting W. Halliday in *Forum World Features*, London, during the autumn of 1971 under construction throughout the world were 330 ships each over 100,000 tons, 183 between 200,000 and 300,000, 10 over 300,000, and one of 477,000 tons. Now these have all been launched and are sailing, and plans for even larger ships, from 750,000 to 1,000,000 tons, are on the drawing boards.

While these huge vessels are able to transport petroleum at reasonable cost, they are not without serious problems. Several somewhat strange explosions have taken place in them and ships have sunk, one as late as November 5, 1973. The cause of the explosions is thought to be related to the large size of individual storage tanks, and the design of the vessels now being built is considerably altered from that of the original large tankers. The oil tanker is relatively fragile and likely to be damaged if stranded on a sand bar, and not uncommonly the plates and frames of a tanker buckle in heavy weather. At a top speed of sixteen or seventeen knots, a 300,000-ton loaded tanker requires three or four miles to stop even when it uses all methods of braking.

As larger ships come into service, the navigation charts become less adequate. Many harbors and passages cannot be used. *Atlas* gives as an example the Dover Straits between Great Britain and the continent of Europe. The maximum permissible draft in the Straits of Dover is 75 to 80 feet, but a fully loaded tanker of 300,000 tons draws at least 75 feet. Imagine the problems created if one of these tankers should be so unfortunate as to run aground or break up and spill its cargo of oil over the ocean.

The size of the new tankers creates a problem in the size of ports. In the United States ports able to accommodate the larger tankers have not been constructed because of alleged harm to the environment and fear of oil spills.

One of the principal beneficiaries from the closure of the Suez Canal and the resulting construction and use of large tankers was probably South Africa, where vessels stop for stores and repairs. Other nations clearly lost: Egypt, for example, by no longer collecting toll from passages through the Canal, Sudan by the necessity to transport all ocean freight around Africa, and Russia by the added distance its ships had to travel. From Batum, at the eastern end of the Black Sea, it is 4,418 miles to Bombay, India, by way of the Suez Canal, but 12,063 miles by way of the Cape of Good Hope. Although Russia is known to be maintaining a strong naval fleet in the Indian Ocean, its trade in that area must have suffered.

The reopening of the Suez Canal could cause changes in the world. It would lessen the dependency of Europe on supplies of oil and natural gas from Libya, Algeria, and Nigeria, countries that are at present in excellent bargaining positions and making the most of their opportunities. The Suez Canal, however, cannot accommodate the size of the large tankers.

With resolution of its war with Israel, Egypt in early 1974 announced its intention to build a pipeline paralleling the Canal and running from Suez to the Mediterranean. What effect will this have on politics and economics internationally?

The easiest and cheapest method of transporting oil and natural gas is by pipelines. Pipelines play vital roles in both the production of crude oil and natural gas and the marketing of the refined products. Gathering and trunk lines move the crude oil from the fields directly to refineries, or to tankers and barges for shipment to refineries by water. Product pipelines transport the lighter refined products, such as gasoline, jet fuels, kerosene, and diesel fuels, and the lighter distillate heating oils, from refineries or marine terminals to marketing areas for distribution to the ultimate consumers.

According to the American Gas Association there were 828,270 miles of gas pipeline in the United States at the end of 1967. An additional 216,270 miles of pipeline not used for natural gas were in service in the United States at the same time. The network of pipelines is being extended annually, not only in the United States but throughout the world.

The use of a pipeline to convey oil and gas from recently discovered fields along the northern Arctic slope of Alaska to industrialized areas in the temperate zone has been the cause of much heated debate. On one side, there is no question that the oil and gas are badly needed in the lower United States, and in addition the oil companies require some return from the large sums they have invested in exploration in Alaska. On the other side, conservationists have feared the effects a pipeline might have on one of our few remaining wilderness areas.

At first other possibilities were examined. The icebreaker-tanker SS *Manhattan*, financed by Humble Oil Company, probed the feasibility of shipping oil directly around the northern extreme of North America to the east coast of the United States, but although the *Manhattan* completed a round trip successfully, the difficulties it encountered indicated that this is not an economical or practical way to transport oil (*66*). Other proposals suggested the use of a fleet of submarines or submarine tankers to go under the ice pack, or of freight blimps, but in all cases the costs appeared prohibitive.

In 1974 the construction of a pipeline from the North Slope oil fields southward across Alaska to the ice-free port of Valdez on the Gulf of Alaska was approved by the United States government. Two routes have been studied. The alternative would have gone southward along the Mackenzie River in Canada, where it could be tied into the trans-Canada pipeline system and thus require no ocean shipment. If a second pipeline for the transmission of natural gas were constructed through Canada, the gas could be piped directly to areas of consumption without the necessity of building a plant to refrigerate the gas. By the alternative route, oil and natural gas could be transferred directly into pipelines in and around Chicago. However, recent difficulties with the Canadian government concerning shipments of gas and oil to the United States indicate that there would have to be extended and complex negotiations before a pipeline up the Mackenzie River could be constructed.

Approval of construction of the pipeline to connect North Slope oil fields with the coast at Valdez was long delayed, primarily because of efforts of conservation groups to prevent this use of the wilderness. Part of the terrain to be traversed is frozen tundra, or "permafrost"—permanently frozen ground. In order to pump oil through a pipeline where outside temperatures are far below freezing, the oil must be kept warm to reduce its viscosity and permit it to flow; a temperature of 150°F is commonly mentioned. Conservationists are afraid that at this temperature the ice in the permafrost would melt and form a "canal" around the pipeline, which would then settle, so that ultimately the pipe would bend and break, pouring oil on the ground. This would without doubt seriously affect the fragile Arctic ecology.

Conservationists also claim that construction as proposed will disrupt caribou migration routes, resulting in diminishment of the herds. In addition they fear that the pipeline will cross known earthquake belts and thus chances of breakage, with consequent flow of oil over the surface, will be too great a risk. They hold that far greater damage would be done to delicate Arctic vegetation and animal life than would result from a similar break in temperate climates.

The oil companies claim that to the middle of 1971 alone they had spent millions of man hours and more than thirty-five million dollars studying problems related to the pipeline. More than a million dollars was devoted to study only of the ecology (*59*). To avoid thawing the permafrost, the oil companies propose to construct part of the pipeline above ground, although this will considerably increase the costs of installation and maintenance. They also claim that danger of breakage will now be minimal.

No one can feel happy about intrusion upon the wilderness. It is justified only by the urgency of the need. What is its underlying cause but the increase in population? Without a limit on population, how can man keep from infringing on all other forms of life and from overrunning all parts of the earth, even the most remote? At least in this case five years of study and discussion preceded approval of the pipeline, and its construction is certain to be under the eyes of many who care about the Arctic.

Problems of governmental jurisdiction also preceded approval of the pipeline, since most of the northern and central parts of Alaska are federal- and state-owned, some three hundred seventy-five million acres being owned solely by the federal government.

The policy of the state government at times ran counter to that established by the Department of the Interior, apparently the federal bureau best equipped to administer the area. The Department of Defense claimed final jurisdiction over any plans approved by other agencies. The federal courts delayed construction while they deliberated. And, to further complicate the situation, various Indian tribes asserted that the land to be traversed was theirs.

Congress agreed to give the native tribes a billion dollars, to be paid five hundred million in cash and five hundred million from oil royalties, and a clear title to about forty million acres of land, thus apparently settling the Indian question to the satisfaction of all concerned. Alaska proposed that the state rather than the oil companies build the pipeline, which it would then lease to the companies, thereby obtaining revenue.

Studies of the problems continue. The Department of the Interior has completed an exhaustive survey of the two pipeline routes. Eventually both will probably be constructed, because of Canadian and United States demands for petroleum and natural gas.

10

Energy II

~~~~~~

Coal

Like petroleum and natural gas, coal is a fossil fuel. It is the former vegetation of fresh-water swamps which became buried. In the course of millions of years this vegetation was compacted by pressure to form the solid layers that are mined in various parts of the earth. Where the vegetation was partly compacted but subjected to insufficient heat and pressure to change the organic matter to coal, peat was formed. Peat is an inefficient fuel, although in Denmark, Ireland, and some other places considerable quantities are used.

Coal is commonly classified as anthracite (hard) coal and the soft coals that include bituminous, sub-bituminous, lignite, and brown. We have already noted that coking coal, essential in the smelting of iron and steel, is the form of bituminous coal that is low in gas content and high in fixed carbon. The precise scientific classification of coals is based upon chemical composition and requires closely controlled analyses.

During the early part of the present century coal supplied by far the largest amount of the energy used by mankind. Even as late as 1920, coal furnished over two thirds of the energy used in the United States, and in western Germany it was the dominant form of energy until the 1960's. With the advent of the cleaner, generally cheaper, and more easily handled petroleum and natural gas, the part played by coal has gradually diminished. Today only a small percentage of the energy that supplies household and commercial heating comes from coal, and essentially no coal is used in the transportation industry. Only in the electric power industry, where coal is burned as a fuel for electric generators, and in the smelting of iron and steel has it maintained its former stature. About 20% of the total energy used in the United States in 1970 came from coal. In England, however, coal still supplies about two thirds of the energy consumed.

The use of coal in the United States has been strongly affected by recently established air quality standards. Much of the coal available, especially in the eastern and midwestern parts of the country, contains unacceptably high amounts of sulfur, and, while much has been done to remove the sulfur from coal and from the smoke where coal is burned, most coal-fired power plants do not as yet meet requirements.

Billions of tons of low-sulfur coal exist in the western United States, but their use in power plants in the east and midwest has been handicapped by high transportation costs.

Coal used in power plants in the recent past has had to compete with the cleaner, more versatile, and cheaper low-sulfur residual fuel oil and with uranium, since nuclear power is beginning to take a small share of the field formerly held by coal. Now that imports of crude oil are curtailed and prices greatly increased, coal is once again competitive. Its production is being rapidly increased.

Where coal is found. Coal is widely distributed throughout the world, but, as with practically all mineral resources, there are areas of surplus and areas of deficiency. The United States has within its boundaries ample reserves of many kinds of coal which should last for a century or two. The country has been wasteful of its better coking coals; it has used them when noncoking coals would have sufficed and exported large amounts. Coking coals are not overly abundant. Coals other than the coking varieties are effective in power plants and boilers but cannot be used in iron and steel smelting.

Europe contains sizeable deposits of excellent coking coals in the Ruhr and Saar regions of western Germany and in Poland. A few other places in Europe as well as Great Britain have in the past yielded large tonnages of coal, including the coking varieties, but while a few of these old districts still produce small quantities and new coal beds are occasionally discovered, many have passed their peaks of production; the remaining coal seams are thin, deep in the ground, and therefore costly to mine.

Russia and China have very large and excellent reserves of both coking and noncoking coals. So has Australia, particularly in Queensland and New South Wales. The Australian deposits are being actively developed and mined and much of their output is shipped to Japan, which has only a limited supply of low-grade coal. Smaller deposits, upon which local iron and steel industries are based, are mined in India and South Africa, although in much of India and

Fig. 10-1. Coal production in 1970 (anthracite, bituminous, and lignite). "Others" include Czechoslovakia, 3.7% India, 2.6%; Australia, 2.5%; South Africa, 1.9%; France, 1.3%; Japan, 1.3%; Bulgaria, 1%; Rumania, 1%; miscellaneous sources, 6.6%. Total production was 3,288,197,000 short tons. Source: United States Bureau of Mines.

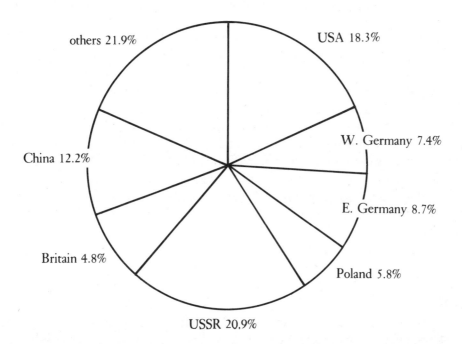

central Africa there are areas of deficiency. South America likewise has insufficient coal for its needs, and many of its coals are low in grade, high in ash content, and expensive to mine and transport; its coal-dependent industries, especially the smelting of iron and steel, require the importation of high-grade coking coals, which are blended with lower-grade domestic coals to form a satisfactory material for the smelting of iron ores.

Coal at present. With modern machinery and the advance of technical knowledge, coal mining methods have changed a great deal in recent years. Big machinery using diesel oil can move large quantities of dirt so cheaply that it is economically feasible to mine from the surface even if hundreds of feet of overburden must be removed in order to get to the coal. No longer does a single coal miner work with pick and shovel. Large machines not only move dirt but cut out and move coal. The mines least expensive and easiest to operate are open pits or strip mines, where surficial soil and rock above the coal are scraped off and piled to one side. An underground mine is limited by the size of its shaft, which determines the amount of output. There is no such limitation on a strip mine, where shovels can be installed at any time to obtain more coal rapidly. In addition, strip mines are far safer than those underground, and little coal is lost in surface operations whereas 50% recovery underground is reasonable. Deep mining remains dirty and dangerous. Few young laborers will want to become underground coal miners unless they receive high wages which will increase the cost to the consumer of coal from underground mines.

In the past many strip mines were worked out and abandoned with no effort to rehabilitate the surface, but at present concern about the environment makes evident the need for laws to provide that the surfaces of all abandoned pits be graded and planted with trees or grasses. Laws must also prevent the acid mine waters that commonly form by weathering of the sulfur in old coal mines from being emptied into surface drainage systems. It follows that we as a society must then expect to pay more than in the past for coal from strip mines because of the extra costs of environmental restructuring.

Coal is usually transported in railroad gondola cars, trucks, and ships, being handled in much the same way as any other solid commodity. In recent years, however, some coal has been crushed and mixed with water, when it is called *slurry,* so that it can be pumped through pipelines, like gas, oil, and water. Facing this threat, railroads that depend for much of their revenues upon the transportation of coal, which requires that cars must make the return trip to the mine empty, have made numerous improvements in handling coal in a so-far-successful effort to prevent further pipeline competition.

During the energy shortage of 1973 when the Arabian countries cut off their exports of oil to the United States and demands for coal soared, the industry was somewhat handicapped by a shortage of railroad cars and barges.

The problems that have confronted the coal industry of the United States in the past are being reinacted at present both in eastern Canada and in the Ruhr region of western Europe. Eastern Canada has long depended upon the coals of Nova Scotia for fuel and energy. A few years ago the transcontinental pipelines connecting eastern Canada with the petroleum and natural gas fields of Alberta were completed, and for the first time abundant oil and natural gas were available to eastern Canadian industries. Industries and individuals alike turned away from coal to oil and gas. The result was a gradual decline of coal mining; unemployment rose and the areas dependent upon the mines became generally depressed. It is doubtful that coal mining there will, while natural gas and oil last, reach even the stage of limited prosperity it enjoyed before the introduction of oil and natural gas. Canada is faced with the problem of retraining the coal miners and relocating them and their families.

The situation in Europe is slightly different from that of Canada. The bulk of the oil and natural gas used in western Europe is imported, mostly from countries that have somewhat undependable and unstable governments, notably those of the Middle East and North Africa. European industrialists recognize the advantages of oil and natural gas but have been reluctant to change from coal to sources of energy that are considered unreliable. Furthermore, the coal industry of Europe employs large numbers of miners who would be thrown out of work if the mines were closed; this would cause unwanted political and social-economic upheavals. Although industry had been turning to oil and natural gas for its energy needs, the changeover is likely to continue to be slow in order that the reduction in the employment of miners may be gradual. The displacement of coal may be accelerated, however, as natural gas and oil from the recently discovered North Sea fields become more readily available.

The future of coal. In spite of competition from oil and natural gas (and nuclear power), and despite environmental problems, coal is sought eagerly and new mines are being developed rapidly. Since demands everywhere for additional energy have been so great, the future of coal mining appears brighter than for many years.

No adequate substitute has yet been found for coking coal in the smelting of iron and steel; this market should be secure as long as the supply lasts. Coal, including the lower grades, may be used to manufacture gasoline and similar materials; oil made from coal was widely used before the advent of kerosene, and during World War II Germany obtained much of its gasoline from the low-grade brown coals of Brunswick. Generally, however, coal has proved unsatisfactory for the manufacture of lubricants such as heavy greases.

A workable gasification process is under study to utilize coal in the manufacture of a substitute for natural gas.

Coal and its byproducts are also the basis for a growing chemical industry.

As power needs continue to grow, so also will the use of coal expand; research should result in adequate methods for the reduction of emissions, chiefly sulfur dioxide and particulates caused by combustion. Coal is at present one of the two most abundant sources of energy available to mankind and it will be needed and used by future generations. Coal deposits will remain after petroleum is gone.

The tides

The marine tides appear to be an inexhaustible source of energy. In many parts of the world, such as the shore lines of northern and western Australia, the tidal range is frequently thirty feet or more. Where the water enters narrow estuaries or inlets, why could not dams be built at the mouths? In this way, water passing either into or out of the inlets could be routed through generators or turbines and its energy converted into electricity.

Such a description is an oversimplification. To control the tide would be both difficult and expensive, and because the marine tide is erratic any hydraulic head it could provide would be intermittent. Moreover, there are not many places where an attempt to use the tide for energy would be feasible. Usually the tide drops and rises less than ten feet, and a dam of this height is insufficient to furnish large amounts of power.

In the 1930's, under President Roosevelt, the United States attempted to utilize the tides at Passamaquoddy, Maine. Results were so costly and unproductive that the effort became known as a national boondoggle. Today the French continue to try, with but small success.

No one watching its ebb and flow can doubt that the tide has energy, but no one can fail to classify that source of energy under those that man is unable to harness effectively.

The wind

Windmills have been used for a long time during the history of man. Those of Holland are probably the most famous, although other places, such as Punitaqui in Chile, also have them. Windmills, like waterwheels, are simple in operation and cause no pollution, and they may even seem to present a possible solution to the energy problem if one fails to understand their limitations.

Like the tide, the wind is erratic. In some places and at some times it is strong, while elsewhere and at other times there may be little or none—it blows hot and cold. What would a hurricane do to windmills depended upon as sources of power? Wind power is inefficient. How many windmills would be needed for the operation of a modern factory or apartment house? How much space would they require? Even Holland has turned to other power sources, and the days of the picturesque Dutch windmills are numbered.

The earth's heat

Heat radiates from the depths of the earth and is a possible source of energy if it can be controlled. Such heat is generally diffuse or, in the case of volcanoes, sporadic and unpredictable. Energy from escaping steam and hot waters has been utilized to a modest extent in Italy, New Zealand, Iceland, California, Japan, and Mexico. The investigation of many other areas known to have hot springs continues in the hope of developing constant and clean sources of cheap power. Probably many more hot springs and thermal areas remain to be developed.

In recent years the use of geothermal or natural steam power has advanced. Deep wells have been driven at the Geysers region in northern California, where the capacity in 1974 was reported to be over 400,000 kilowatts. The Pacific Gas and Electric Company, which uses the power generated by the steam, has announced plans that will add 563,000 kilowatts of geothermal power to its system by 1976. This geothermal system would then furnish as much power as Hoover Dam on the Colorado River.

In southern California, in the Salton Sea basin where the existence of hot brines has been known for many years, efforts are continuing to develop them as sources of power. Since the brines contain large quantities of salts that are precipitated when the fluid is cooled, some way must be found to remove the salts, which quickly fill ordinary pipes. At present a method under trial is to pump relatively pure water underground, to return as superheated steam to the surface, where its energy can be used to generate electric power.

Energy from the heat of the interior of the earth should become increasingly useful. However, geothermal possibilities, which exist chiefly in or near volcanic and earthquake belts, are not widely dispersed over the earth.

Uranium

The radioactive mineral uranium, promising to provide abundant inexpensive energy obtained from nuclear reactors, has captured the imagination of energy-hungry people everywhere. But unfortunately its promise has been slow to materialize, and power generated from nuclear reactors is still a minor part of the energy used in the world.

Probably the most common mineral of uranium is an oxide—a chemical compound of a mineral and oxygen—called *uraninite* or *pitchblende,* although many other mineral compounds of uranium are used in the preparation of the fuel for nuclear reactors.

Fission. Though uranium had been known for years, and components from the decomposition of radium had been widely employed for medical purposes, it was not until the Second World War that the potential of nuclear fission was generally recognized, by the discovery of methods of controlling the speed of reaction. One of the isotopes of uranium, uranium 235, was found to be especially susceptible to control; this is the substance that is used in hot water nuclear reactors or "burners." At first its conversion to more stable compounds was accomplished with explosive force, in the atomic bomb, but gradually methods of slowing and closely controlling continuous conversion were developed. The peaceful use of atomic reactions became a reality, and will perhaps be of major importance.

Radioactive decay, or the fission of atomic nuclei, chiefly in uranium and thorium, produces radioactive byproducts and wastes. Radioactive byproducts are finding use in medicine and industry. The wastes stabilize so slowly that some remain hazardous for centuries and others for thousands of years. It is the bombardment of life by emanations from the fractured particles that causes so many problems. Radioactive wastes are dangerous to the whole environment, including all forms of life. If they are not properly stored or neutralized, they may cause extensive harm by getting into water supplies.

So far their disposal is an unsolved problem. In the past the wastes were dumped in deeper parts of the ocean in steel and concrete drums or buried underground in steel tanks. Proposals suggest burying them in closed drainage basins or in impermeable areas such as massive salt beds where the circulation of ground water can be controlled, thus preventing leakage. The peril of accidents in the disposal of the wastes remains a real one. What containers will be safe for thousands of years? What places will make safe repositories for such a long time?

Nuclear reactors have been severely criticized also because they are very inefficient (utilizing an amount of uranium small in comparison with the amount discarded) and produce large quantities of waste heat. In most reactors disposal of the waste heat is accomplished by the use of cooling water which thereby becomes hot. The heated cooling water was in the past poured into drainage channels near the plants, with resultant damage to the ecology of the areas. Now the newer reactors use cooling towers in which water is recirculated and contamination of the drainage system by heat is largely avoided.

If the world's uranium is to be used in fission reactors like those now in operation and under construction, the supply of the radioactive material, uranium 235, will be exhausted at about the same time as the world's coal supplies. Thus long-term hopes for cheap, clean energy from radioactive minerals in the future lie with a nuclear reactor that operates on a different principle, the so-called breeder reactor.

Uranium as found in nature is 99.28% uranium 238, which is not radioactive; the remaining $^{72}/_{100}$ of 1% is the radioactive substance, uranium 235. By bombarding uranium 238 with neutrons from contained uranium 235, a reaction can be established that converts the otherwise useless uranium 238 to plutonium 239. Plutonium 239 is fissionable, like uranium 235, so in this way part of uranium 238 can be converted to energy.

At present there is no commercial breeder reactor in operation in the western world. Russia claims to be operating one. The United States has built two "small" experimental breeder reactors that have demonstrated the feasibility of the operation.

There are two hurdles to the realization of the hope offered by the breeder reactor. One is the technological problem of building a safe furnace. The other is the speed of breeding of radioactive material, a speed controlled by nature and apparently not subject to the control of man. Will there be sufficient time to permit the construction of enough breeder reactors? Several decades may pass before they come into commercial use. Still, a government spokesman declares that by the year 2000 the United States will need about three hundred breeder reactors to satisfy its energy demands.

Uranium mining. The mining of uranium ores has been conducted in a normal fashion and was thought to be safe from radiation hazards until a few years ago, when it was determined that radon, a radioactive gas, existed in underground mines in quantities dangerous to health if workmen were exposed for long, continuous periods of time. Radon is the residue left after radium disintegrates and loses alpha particles. Large amounts of fresh air must be circulated through the underground workings where radon is detected, and the permissible standards for this gas have been established high enough to make the underground mining of uranium difficult and generally unattractive from an economic point of view. Open pit mining is thus greatly preferred.

Where uranium is found. Like other natural resources, uranium is scattered irregularly over the earth. Most of the economic deposits are in sandstone, although a few are known in limestone, shale, and other kinds of rock. Large quantities of the element are recognized in black shale and coal, but the percentage of uranium in these materials is too low in grade to permit profitable mining under present economic conditions.

The United States is lacking in high-grade uranium ore. Its moderately large reserves of low-grade ore are insufficient to permit complacency, so exploration needs to be continued in order to discover new deposits. The deposits of northern New Mexico near Ambrosia Lake and of several districts in Wyoming are the largest and most productive in the United States. Smaller amounts of ore are in Texas, Colorado, Utah, South Dakota, and some other states.

The best known and largest deposits of uranium ore in other parts of the world are in Australia, Canada, South Africa, and Russia; some of these contain ore of much higher grade than that in the United States. Small deposits of uranium ore are recognized in numerous other localities, but some regions, notably western Europe and South America, which are deficient in oil and natural gas appear to be deficient in uranium as well.

Uranium in the future. In 1970, nuclear-generated electric power supplied no more than about two tenths of one percent of the energy consumed in the

United States (*12*), but at present construction proceeds on several new power plants badly needed to supply the demands of industry. Even so, nuclear reactors are not being built as fast as was predicted only a few years ago. It is now estimated that by 1980 they will provide less than ten percent of the energy required in the United States.

The development of nuclear reactors has been retarded by their complex technology, more difficult than had been expected, and by opposition on the part of the public. In spite of nearly perfect safety records, people are still unwilling to accept nuclear reactors near homes, fearing leakage of radiation and feeling concern about the disposal of radioactive wastes.

Nevertheless, energy from uranium is needed to supplement that from other mineral fuels, it is readily available, and the process of recovery has been proved to be economically feasible. In spite of the problems it presents, this source of energy may in the future be the solution of part of the world's need for power.

Thorium

Thorium is another radioactive element that is widespread and may eventually turn out to be useful in providing atomic energy. The minerals of thorium are even more difficult and expensive to process in the production of fission reactions than are those of uranium, but they are found in deposits in many places. Thorium in the mineral monazite frequently accumulates in beach sands, where it may be cheaply recovered. As Chapter 4 notes in its discussion of titanium, the governments of both India and Brazil have sharply restricted what were formerly thriving industries of mining beach sands, in order to conserve the thorium content in the hope that thorium will prove to be an economic source of power.

Thorium 232, the most abundant kind of thorium, is only mildly radioactive. However, when thorium 232 is bombarded with neutrons, it eventually changes into the fissionable uranium 233.

As we have pointed out, the breeder reactor—which uses uranium 233 made from thorium 232, or plutonium 239 made from uranium 238—could greatly expand the world's supply of nuclear energy.

Hydrogen

Hydrogen and oxygen are the two components of water. If water can be cheaply broken into its components, the hydrogen would make a clean and useful fuel. When burned, hydrogen recombines with oxygen and forms water, giving off heat in the process. These are not radioactive materials, so no pollution results; but hydrogen is highly explosive and must be handled with great care. However, for the process to be effective, less energy must be required in the release of hydrogen from water than is given off during its recombination with oxygen.

Preliminary research indicates that the bombardment of water with a stream of neutrons to yield hydrogen deserves further investigation. It may turn out to be a useful energy source.

Fusion. Fusion is a nuclear possibility different from fission. The fusion of the isotopes of hydrogen, found in what is called "heavy water," deuterium and tritium, into helium is the basis of the hydrogen or thermonuclear bomb. In this reaction the energy is released with explosive violence in tremendous amounts. Studies of means of controlling the speed of reaction have been carried on in many places, but so far without much success, principally because of the extremely high temperatures, millions of degrees, required. Such temperatures would vaporize any known furnace. It is possible that plasma physics may hold the answer to the problem. The principle is that of an electromagnetic field that would "hold" and control the fusing gases as no standard material could.

The sun

All of the energy used by man, except that radiated from the interior of the earth or by the decomposition of uranium and other radioactive unstable elements, comes from the sun. Sunlight is inexhaustible, but we do not yet know how to harvest or to store the energy from its diffuse rays. To date, solar energy is available to man only indirectly, through plants, which, by the process known as photosynthesis, use it to produce their body substances. Because human beings eat plants—as well as other animals that also eat plants—all of our human energy comes indirectly from the sun.

Petroleum, natural gas, and coal are believed to have originated from the decomposition of ancient plant matter, and thus the energy locked up in these fuels has come directly from the sun. "Liquid sunshine" is a term often applied to petroleum.

Energy in the future

All through our discussions of nonrenewable resources you must have noticed that the decisive element in every aspect of man's relationship to the earth is energy. The extraction of metals, mining at deeper levels, using lower grades of ore, ocean mining, recycling, plastics and other substitutes for raw materials, technology, maintaining a clean environment—all depend on energy.

Fig. 10-2. Projected energy requirements per year in the United States, in quadrillions of British Thermal Units. The projections from 1970 on, shown by the dotted line, are based on a doubling time of ten years, while the solid line shows a doubling time of twenty-five years. Source: Cordell Durrell.

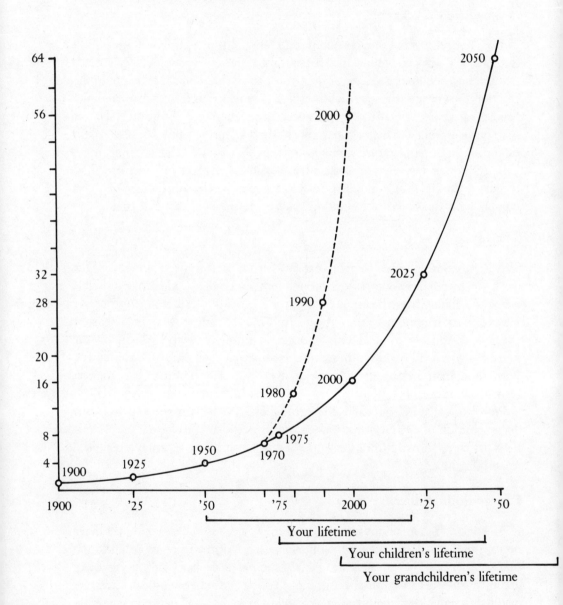

Some nations now possess or can afford to buy sufficient energy for their needs—their relative wealth and better standards of living show the result. On the other hand, in most of Africa and Latin America many commodities are expensive and wages and living standards are low because nations must rely on manual energy, being too poor to buy any other kind.

If available energy becomes more expensive, standards of living must fall everywhere because the costs of all commodities will rise. As some nations seek to increase their revenues by raising the price of the petroleum they export, they increase costs in the industrialized nations, costs not only of petroleum but of other commodities. Such increased costs, if excessive, will eventually lower the standard of living in the petroleum-exporting nations as well as in the countries that must import.

That is the paradox that civilization faces today. Living standards are set by the amount of clean, inexpensive energy available per capita. These three aspects of energy—its abundance, cleanliness, and cost—interact upon each other. As more energy is used, less is ultimately available and greater pollution results. Yet the fight against pollution increases energy costs and hinders the development of energy sources such as those of Alaska's North Slope, nuclear reactors, and coal. Costs of traditional sources of energy are rising principally because of the wish to improve living standards. Yet such rising costs, if they continue, must result in lower living standards.

Review of the sources of energy. The principal sources of energy for the world today are the fossil fuels—petroleum, natural gas, and coal. Being mineral resources, they are subject to the natural laws that govern all such resources. They are finite in amounts. They are unevenly distributed throughout the world. They are exhaustible; the petroleum pool that takes 100,000 years to form can be depleted in a man's lifetime.

At present fossil fuels are available in sizeable quantities and from many suppliers. However, the prices of petroleum products to the ultimate consumers have not only risen recently but are expected to rise further in accordance with the announced policies of the major proprietary nations. Many of the developing countries that produce petroleum for sale on the world markets are exerting more and more controls over production as well as insisting on higher prices.

The need for maintaining low-cost energy is not generally recognized even in the consumer nations, where gasoline and fuel oils are convenient commodities for taxation because the taxes are easily collected. For example, the State of California, in the fall of 1971, added 2¢ a gallon to its gasoline tax to support public transportation development. In California anyone buying a gallon of this essential fuel was thus paying a federal tax of 4¢ and a state tax of

7¢. Then in July of 1972 California added a 5% sales tax on the retail price of gasoline *including* federal and state tax, although the tax on tax was later rescinded. While a consumer is not everywhere forced to pay tax on taxes, taxes always constitute a major part of the cost of gasoline to the consumer, and there always exists a pressure for their increase.

Fig. 10-3. Forecast of the United States petroleum industry's capital requirements for expansion of production facilities to meet increasing demand. Source: *Seventy Six*, 1972, Union Oil Company.

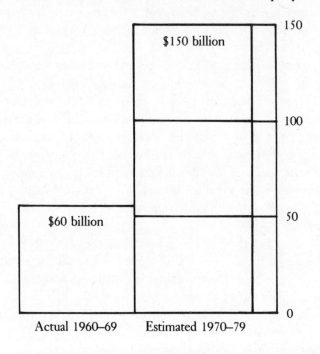

Actual 1960–69   Estimated 1970–79

Numerous reports have been written about the amounts of fossil fuels available in the world. In fact, probably more attention has been paid to the fossil fuels, especially gas and oil, than to any other raw material (*31, 41, 51, 32, 33*). Shortages have been predicted for decades. They have failed to materialize because of the periodic discovery of new oil or gas fields or coal deposits, or the development of a substitute, or an improvement in technology. Still, at no time in the past has the world been faced with the tremendous population growth that has occurred during the last few years, and at present some authorities are predicting approaching exhaustion. Others claim that sufficient mineral fuels exist for most purposes for many years and that further exploration will reveal large additional supplies.

All authorities seem to agree, however, that at some time in the future, possibly shortly before or after the turn of the century, insufficient organic fuels will exist to smelt the great amounts of low-grade ore that will be needed and also to fulfill the other requirements of humanity. There is no question but that the earth's fossil energy will run out, if not suddenly, at least over a period of years. Industrial centers are likely to feel the pressures of shortages first, because industry is the large user of cheap energy, yet every individual, as well as every nation, will be affected by the results, simply because every one of us uses the products of industry. The United States, for example, is vulnerable, as it currently consumes about 33% of the world's petroleum production and has only about 6% of the world's reserves (*9, 64, 7*).

The fossil fuels are not clean in the ways in which we use them to produce energy. As we burn them, byproducts pollute either the atmosphere or the environment and adversely affect the quality of life. Nevertheless, until other supplies can be developed, the consumption of energy from the fossil fuels is bound to increase unless civilization stagnates or declines. So long as population continues to grow and standards of living to improve, the demands for energy will grow.

An excellent summary of the situation faced by the United States was given in 1970 (*8*); since then it has intensified:

Suddenly, after many years of abundance of energy, we find ourselves with insufficient supplies of natural gas, no economic spare coal production capacity, inadequate reserves of electric generating capacity, insufficient supplies of clean fuel to meet the regulations governing air conservation, and a rapidly increasing dependence upon foreign oil.

The chief sources of clean energy are falling water, the tides, sunlight, heat from the earth's interior, and atomic fusion. Falling water for hydroelectric plants is limited by the amount of rainfall, a fairly constant annual figure for the world, and most rivers near the industrial centers have already been harnessed. The tides have not been widely utilized in spite of many attempts, and neither has sunlight, to any appreciable extent. Neither has the earth's heat, except for a few moderately promising geothermal projects. Atomic fusion requires a temperature of millions of degrees, and so far there has been no answer to the problem of designing a furnace that will not vaporize at such a temperature.

Nuclear fission means risk to our environment, chiefly from the disposal of atomic wastes or cinders, dangerous to ground water, dangerous for thousands of years. We do not know all that there is to know about the radioactive elements, especially the technology of handling them.

What now? Because the fossil fuels are exhaustible and already past their peaks in available supplies, they must be used carefully to tide us over until we can develop other sources of energy. The possible sources appear to be:

• Nuclear reactors, using uranium, thorium, and plutonium. We shall probably have to develop them, in spite of their technological problems and their environmental risks. These, especially the breeder reactor, are perhaps the best bet for the next two decades.

• The burning of hydrogen. The process of obtaining hydrogen from water would be ideal in providing a clean fuel. The technology of burning the hydrogen, once obtained, would be simple. The difficulty lies in obtaining the hydrogen by releasing it from water at a low cost in energy.

• The sun. Just because solar energy has not been harnessed successfully by scientists working for almost a century does not mean that it cannot be done. Sunlight, like hydrogen from water, would provide us with clean energy. In addition, sunlight is inexhaustible. As an early solution, this source is more reasonable if not more likely than the burning of hydrogen.

• The fusion of hydrogen into helium. Balancing on one side the inexhaustibility of hydrogen, the scientific understanding of the process, and the fact that pollution is avoided with, on the other side, the terrible complexity of the technological problems leads to the conclusion that this may be the greatest possibility for the next century.

We must also work toward conservation measures such as the development of a workable fuel cell. This would permit the storage for later use of electricity that could be obtained at times when generators were operating below capacity, and would thus tend to level out the peaks and valleys of consumption.

None of the possible new sources of energy appear easy to develop. But the alternative will be the return of the world's societies to a small population with a lower standard of living. Once again man would become, as he once was, tightly tied to small patches of soil where at best he could exact a meagre subsistence with long hours of back-breaking labor.

Will man in the future obtain adequate amounts of energy to maintain his civilization? Will energy in the future be cheap enough so that the average person everywhere can afford to use it, and clean enough to satisfy the exacting environmental demands of a crowded world? These questions are man's greatest burden and his greatest challenge. We do not know the answers. They depend upon man's technological ingenuity and his common sense (*52*).

11

Soils and Water

Soils

We do not cherish what we "treat like dirt," but if there is anything we ought to cherish, it is our soil. We live on it. It gives us our nourishment, our energy.

The average modern city dweller, whose feet may not touch the ground for days on end, is so insulated from the soil that he scarcely realizes that his life comes from it. Combined with the energy from the sun, soils are the basis of plant life, which in turn is the basis of life for all animals, including man. Man cannot obtain from the sun the energy he needs except through eating plants, or animals that have eaten plants. By the process of photosynthesis, plants utilize energy radiated from the sun to convert water, carbon dioxide, and minerals in the soil and in fertilizers into starch, sugar, fats, and other substances. By consumption of the plant products animals and men, through the combustion process of digestion, obtain the sun's energy. As an oversimplification we might say that here in the soil are the minerals that nourished the plants that utilized the sun's rays to give Jack the energy to build his house or to do anything else, physical or mental.

What soils are. There are different definitions of soil. To the agriculturalist it is the loose accumulation at the earth's surface, including organic matter. To the engineer it is any of the earth's crust that can be moved without blasting. To the geologist it is an unconsolidated decayed portion of the earth's inorganic crust. To the geologist concerned with water supplies, the beds of the crust that carry water may be either soils or porous rock strata known as *aquifers*.

Soil is created over a long period of time as rock, a solid mineral mass, gradually decays to assume a loosely granulated form. It may lie directly over its parent rock or it may have been transported by wind or ice or water to accumulate in sheltered areas. It may be relatively homogeneous or a heterogeneous mineral conglomeration of many rocks that have been weathered. The various minerals which are, one might say, frozen in the different kinds of rock are released into the common mass, soil, which forms the loose layer at the earth's surface.

Eventually soil is carried to the sea and buried. With pressure, chemical action, and heat, it again becomes rock, completing a cycle that may span thousands of years in which minerals were set free and gathered up once more into a solid mass.

All soils are part of this cycle, and all rocks. The wind that blows soil away and the rain that washes it are also slowly weathering rock into soil. The process, like everything done by nature, does not work on a uniform basis. In places the agencies of weather act much faster in carrying off soil than in making soil from rock, and when man anywhere finds that his soil is disappearing, he faces trouble.

Soils and weather. In the mid-1930's strong winds blowing across the southwestern United States made what was called the Dust Bowl out of the Great Plains as, in unusually dry weather, they carried off the soil. Unable to continue growing crops, the people migrated in large numbers and became the very symbol of poverty. The wind had made them poor by taking away their soil—an estimated several hundred million tons of it. Many eons will pass before it can be replaced by normal processes of weathering.

In the tropics there are places where heavy rainfall has leached, or washed out, the nutrient minerals from the soil. Soluble minerals and organic nutrients, as well as many of the minor and trace elements needed in small amounts, are gone, and the crops produced are very poor; they may appear lush and green, but they lack the minerals necessary to nourish animal life. In places the soil is reduced to small pellets and the alumina- or iron-rich clay known as laterite, a potential source of nickel and aluminum but in general incapable of maintaining agriculture. A visitor from Iowa may find the soils of the tropics, despite the appearance of their growth, far less productive than those of his native state. As yet there has been no satisfactory answer to a question of growing importance: Can large areas of the tropics, underlain by laterite and subjected to torrential rainfall, be made to feed large numbers of people?

Under conditions of this kind, people have to find substitutes for soil crops. They hunt or depend upon food from the water and from deep-rooted trees—upon fish, nuts, and various fruits. Fertilizers and soil conditioners can be used to improve the soil, but the amounts needed to make normal farming feasible in even a small part of the Amazon River basin, for example, would be prohibitively expensive. Correction of soil deficiencies in the temperate zones is a matter of routine treatment, but in areas of heavy tropical rainfall this is generally impractical. Even in areas of the rain forest where soils are protected by thick covers of vegetation, the soluble constituents essential to plant nutrition are rapidly washed away.

In some places soil has not yet suffered from leaching despite heavy rains. In Hawaii the geologically recent lavas yield soils very rich in minerals. Where the rainfall is heavy in the Hawaiian Islands, on the summit of Kauai for example, the mineral content derived from the decaying lavas replenishes what has been dissolved.

In addition to wind and rain, ice strongly affects the soil. There is no soil on some rocky areas where glacial ice has scraped away everything loose, to leave only the bare rock.

Variety of soil. Since any rock may be the source of soil, and since rocks vary greatly in their mineral composition, there are tremendous variations in the mineral content of soils. There are places in the world where cattle may starve because the grass they eat has, through lack of mineral nutrients in the soil, a corresponding lack of nutritious value. Cattle may die after eating plants that have absorbed the selenium from some soils, in parts of the Dakotas for instance. On the delta of the Amazon, tall grass has the cutting power of glass because of the high percentage of soluble silica in the soil in which it grows.

Insoluble constituents such as alumina and the oxides of iron tend to remain and to concentrate. These are the end products of all soils.

Care of the land. Although agriculture is as old as man himself and has always been valuable to him, man has often treated his soil as if it were dirt-cheap. He has allowed erosion to remove the topsoil (the richest soil at the surface). Where gullying, the wearing away of the soil so that gullies or cuts appear, has been permitted to get out of hand, as it has in many parts of the world, badlands and rocky deserts develop. The land then supports little if any vegetation and loses much of its usefulness to man (27). Land from which the soil has been removed can generally be restored, if at all, only with great difficulty and over long periods of time.

A convenient though by no means unique example of what might be termed "backward" agriculture is seen in parts of southern Chile. Here the broad rich valleys are planted in the slowly maturing but highly prized pine tree of the region. These trees have been treated as cash crops, generally by absentee landlords. Meanwhile, the steep slopes that would readily support a stand of trees are the pitiful grain fields of the poorer people. The result is what might be expected, soil erosion on the cleared and cultivated hillsides and choking of the rich valley floors. Similar or worse conditions are found in many parts of the world, for example in the Middle East and parts of North Africa where good crops once flourished. Because the destruction of soil constitutes a waste of national resources, care of the soil should be the direct responsibility of governments. It is true that eroded areas can be reclaimed if erosion is stopped soon enough; in places this has been done through the long years of expensive, discouraging work that are required before denuded land can again become productive.

Where erosion has gone so far as to remove nearly all soil, leaving useless and almost inaccessible badlands, little reclamation is possible. This is the case in some areas of the United States in which smelters were built and operated around the end of the 19th century. At Ducktown, Tennessee, in the middle of the Appalachian mountains, surrounded by heavy green vegetation, and at Northport, Washington, there are places where the sulfur fumes from smelters killed all plant life and the soil cover was allowed to wash away unchecked. Smelter operators have long since learned how to recover and use the fumes, but years of costly effort have gone into the planting of trees and other measures to help restore beauty and greenery which have been only in part successful. After fifty or sixty years, these areas are still largely desolate, deeply gullied, and unsuitable for any purpose whatsoever.

The need for erosion control is evident in other places, for instance the state of Oaxaca in Mexico and Haiti, where steep slopes have been washed into the sea, leaving only partially decomposed rock instead of soil. The mineral nutrients are there, but they remain locked in the rock, which is not weathered enough to make them available to plants.

Soils are closely related to economics and politics because of the importance of food. In general, countries with rich soils are prosperous; where the soils are thin and poor, man may survive, but life is not easy. Since one of the critical problems in the world today is the feeding of the burgeoning population, every bit of arable land should be preserved by good agricultural practices, which can greatly increase the productivity of soils. Contour plowing reduces the harmful effects of erosion, crop rotation conserves and helps restore the minerals in the soil, while fertilizer and soil conditioners, as we have seen, supplement them and can change poor crops into good ones. Modern agriculture has made great technical advances in recent years. Many governments maintain soil analysis services to advise farmers of what is needed to correct deficiencies in soil content. It may be that a variation in method of irrigation has changed the supply of minerals in the soil. Sometimes all that is needed is the addition of a very small amount of some mineral that has been washed away by irrigation or by rain, or that is required by a certain kind of crop. The health of a tree, and its crop, may depend on the addition of a mineral that is needed only as a millionth part of a leaf or fruit.

In any case, and although too much of it can be so harmful, if the mineral nutrients in the soil are to be available to plants, one thing in addition to sunlight is essential—water.

Water

By dissolving and disseminating mineral nutrients in the soil, water maintains plant as well as animal life. In spite of the fact that it is necessary for all life, water is, like soil, generally taken for granted today in industrialized countries, where people use it with little thought of conservation.

The per capita human consumption of water depends upon the requirements of agriculture, industry, and domestic use. People in the United States use water more lavishly than do people in other parts of the world. Consumption in the United States averages between 150 and 160 gallons per capita per day, while in the major cities of Europe daily per capita consumption averages only 70 to 80 gallons. Elsewhere in the world consumption is far less. Few places are able to afford the luxury of 30 to 60 gallons for each shower, 5 to 7 gallons each time a toilet is flushed, and 17 gallons per cycle of a washing machine. For the majority of mankind, the water service enjoyed in the United States is still a dream. This is exemplified by the case of a Parisian who made frequent trips to New York; every time, upon arrival at his hotel, the first thing he did was put his face under the faucet and imbibe deeply, even though the water tasted strongly of chlorine.

It is estimated that the use of water in the United States will double in the next twenty years. Most of this increase is expected to be in the water used by industry.

The control and conservation of water. In caring for his land, man learned to conserve water, to build dams and reservoirs for watering his animals and irrigating his crops. He learned to keep water on the soil, to encourage it to sink into the ground and not to allow it to run off rapidly.

In places, soils made productive by irrigation water, as contrasted to rain water, absorb the chemical salts in the irrigation water, which causes the soils eventually to become saline and therefore unproductive, as in parts of North Africa. Even so, irrigation remains a necessity in many areas.

Nowadays the conservation and control of water are especially necessary for three purposes: to water arid but potentially productive lands, to supply thirsty urban and industrial centers, and to prevent destructive and wasteful floods.

An excellent example of an international waterway where every drop is of value for sorely needed irrigation and industry is the Jordan River that flows near the border of Israel and Jordan. How to allot and use this water has been a thorny problem. Israel wants it to irrigate the potentially rich but thirsty Negev Desert in the south of the country; Jordan, where a large part of this water originates, also claims a need for all of it. Yet too little water flows in the Jordan River to fill the demands of either country, let alone both.

This particular problem is further complicated so long as Jordan and its Arab neighbors will not recognize the existence of Israel. Thorough studies have been made by impartial engineers and plans have been recommended to divide the water between the two countries, but Jordan has consistently refused to arbitrate because it claims that such an acceptance would imply political recognition of Israel. Possibly the final solution for both countries will come after cheap processes of desalinization of sea water have been developed. Meanwhile water affects politics around the Jordan, as in other places.

The problems of obtaining sufficient water supplies for urban and industrial centers are constantly becoming more pressing, and if world population continues to grow and industry to expand, the need for more water will become increasingly acute. More than once, after a period of drought in its watershed, New York City rationed water, as did most of the surrounding towns. In Rio de Janeiro many hotels formerly had lines painted inside bathtubs to indicate the maximum amount of water permitted for a bath. Quite a number of city dwellers in the past few decades have had to ask for it if they wanted water in restaurants, and garden watering has been limited in many places during summer months.

These irritating stopgap measures merely emphasize the need for water in urban areas. Los Angeles, which is a particularly large and sprawling urban center, by nature a desert since there is low annual rainfall, uses water that is piped in from the Colorado River, from Owens Valley in the eastern Sierra Nevada, and from almost any other region where water is available to it. Still, shortages persist; the largest earth-fill dam in the world now defaces the scenery near Oroville in northern California so that water from the Feather River may be transported five hundred miles, through a series of canals and reservoirs, to Los Angeles. There has been talk of bringing more from as far away as the Columbia River, twelve hundred miles to the north.

The problems of obtaining adequate water supplies are not restricted to large centers of population but are increasingly critical in many smaller communities. With so many uses for water, as a solvent and as a cleaning or cooling agent, for example, industry can operate only where sufficient quantities of clean water are available.

People everywhere have long considered water to be an inexpensive or even a free commodity. Indeed it has been that, and has been freely used in many places. Now the question is arising of how much people are willing to pay for continued unlimited supplies of water. Will they be willing to use purified sewage and waste waters? Probably the present water supplies in many communities can be used much more effectively and with less waste than at present, but even so, supplies in numerous places will need to be greatly augmented in the near future.

As cities reach out to take water from distant rivers, the water rights of areas near the rivers must be protected, and so must the rivers themselves. Some thought must be given to the conservation of rivers in their natural state—they cannot be utilized for industrial purposes alone without being destroyed as a part of the wild. We should not forget the necessity for the multiple use of this resource, or for the prevention of its contamination. Unfortunately, one has only to drive through the Hudson and Mohawk Valleys of New York or many other places to see rivers that, for all their beauty, are limited in value as a water resource because they are polluted.

Sources of fresh water. Water supplies are obtained from many sources; four are of particular interest here: first, the accumulation of run-off or surplus water in dams and reservoirs; second, the development of underground supplies and wells; third, the use of waste waters and sewage; fourth, the development and use of large-scale desalinization plants.

Dams. Dams are effective to store reserves of water for use during the dry seasons, to create power, and to prevent flood waters from damaging the river valleys. Where warm spring thaws quickly melt snow, or long periods of heavy rainfall cause rivers to reach flood stages, dams can forestall great damage. Where homes are built on the flood plains of rivers, there will be havoc when the rivers flood, as the Mississippi did in the early spring of 1965 and again in 1973, unless dams afford protection.

Flood plains make excellent agricultural land. Some of the best crops in the world are raised in areas that are subjected every year to flooding which infuses the soil with new nutrients. One cannot help but wonder what effects on the agriculture of the Nile River delta the high Aswan Dam will have as it prevents the natural annual flooding. Early reports indicate that not only is the annual deposit of silt and nutrients failing to occur, but erosion is now eating into some of the better farm lands.

Flooding may be controlled by the construction of very large, expensive dams, and such dams can furnish large amounts of energy to industry as well as water for irrigation, but, in addition to being costly, they generally remove sizeable areas of good river-bottom land from cultivation. Can smaller, cheaper dams built on all of the tributaries be as effective for flood control? Or should there be both large and small dams? These questions will have to be answered, to conserve both water and soil and to prevent flood damage.

Wells. Many communities and industries obtain their water from underground supplies through wells, and probably a large percentage of the users do not realize that the amount of water stored underground is finite. It exists in limited amounts and is just as prone to exhaustion as is water stored on the surface. How many communities that use underground water have made studies of their supplies and know how much water can be removed annually without permanent damage? Our coastal areas show numerous examples of the way excessive pumping of fresh water has brought about the inflow of salt water into what were once fresh-water-bearing strata. As a result, land that once yielded good crops now supports only salt grass. Even in the interior of land masses, especially in semiarid sections, excessive pumping has lowered the water levels. True, in time water may be partly recharged if production is curtailed, but this is a slow process and may be inadequate to support the community using the water.

It is also true that in general the soils become compacted after pumping; they settle and will not again hold the former amount of water. In the region near Phoenix, Arizona, where the city has spread over a large area and farming has become extensive, excess pumping has lowered the water level in places to alarming depths. More water is being removed annually than is being put back underground. This situation can go on for some years until water becomes saline, until pumping depths become uneconomic, or until the water is exhausted, but eventually additional water to recharge the underground supply must be found or the use of water will have to be curtailed. In Mexico City excessive draining of the old lake beds beneath the city has caused the land to settle; many buildings in the city have cracked and become unsafe because of the compaction of the soils. Such settling is common in places where over-pumping or deep drainage has occurred.

Waste waters and sewage. Waste waters and sewage are usually purified in treatment plants and then pumped into rivers or other bodies of water where they are dispersed and lost. In a few places waste waters are reused; if necessary they may be pumped back underground where migration through rocks is effective as a mode of additional filtering and purification. This method of forcing waste water underground is also an effective way of recharging heavily pumped water-bearing beds. The thought of using sewage and waste waters is repugnant to some people, but with modern methods of treatment there is no reason why these waters cannot be made safe to drink. As supplies of water become more difficult to obtain, it seems likely that more and more waste water must be utilized, although its recycling will require energy.

Desalinization. In recent years a great deal of publicity has been given to the treatment and desalinization of salt and brackish waters. Such waters are found not only along the sea coasts but also in many deep strata in the interiors of land masses and, as we have seen, in sedimentary basins associated with oil and gas. Processes and devices that effectively remove salts are on the market in several forms and used in many places, generally where small amounts of potable water are needed and costs are a secondary consideration. People in coastal cities have long thought of the availability of sea water, and cities such as Los Angeles have directed considerable research toward desalinization. At present many interior cities and, particularly, potentially rich but semiarid agricultural areas, hard pressed for sufficient water, are also looking at saline springs, deep saline waters, and the edgewaters that border many oil and gas accumulations. A desalinization device was installed at the Guantánamo Naval Base in eastern Cuba at the time Premier Castro shut off the water supply. All water at the base now comes from the ocean.

Is it simply a question of time until desalinization can compete successfully with other supplies of potable water? Costs are still somewhat higher than what people are accustomed to pay for water, but in places water may well be worth higher costs. For example, on the sheep and cattle runs of central Australia water is extremely scarce, and much of it is brackish. Would small desalinization devices work here? If desalinized water is to be conveyed to inland cities there will be the additional cost of the energy that will be required to pump the water from sea level to, for instance, Phoenix or Oklahoma City.

Desalinization of water inland may also create the problem of disposing of the salt that remains as a waste product.

Possibly water will continue to be a cheap commodity in many areas, but in others its cost is bound to increase; these regions are the places where desalinization will first be used on a large scale. A desalinization plant designed to produce more than two million gallons of fresh water daily, plus electric power, has been installed at Key West, Florida. Another, the Clair Engle Desalting Plant at Chula Vista, California, began operations in 1967 to pump a million gallons of water daily to the city of San Diego.

In recent years a great deal of interest in desalinization has been shown by the thirsty nations of the Middle East. In November, 1971, the government of Saudi Arabia announced a combined desalting and electric generating power plant capable of producing five million gallons of fresh water and fifty megawatts of electricity daily. Both water and electricity were designed for use in the Red Sea port of Jeddah and for irrigation of the surrounding area. Two separate units were constructed and the site has space for an additional eight units of similar design (55). This installation should be watched with interest. The Middle East in many places needs water, both for irrigation and for home and industrial uses, and it has surplus gas that can readily be used for fuel. The installation of combined desalinization and electric power plants in this region cannot help but mean considerably improved standards of living.

The initial cost of a desalinization plant is small in comparison with that of a large dam; the installation in Saudi Arabia is said to have cost about twenty million dollars, and its size can be increased simply by adding more units as they are needed. These advantages appear noteworthy when one realizes that the Oroville dam and canal system cost approximately two billion dollars of California taxpayers' money. Such a sum spent principally to permit the transportation of water to a place five hundred miles away represents a big capital investment before a drop of water reaches its destination.

The future

Providing adequate water supplies for growing urban centers is not the only water problem today. The need for water for agricultural purposes is also increasingly critical as the population expands and greater amounts of arable land are sought. Irrigation is often necessary in order to obtain the maximum yield from soils. Irrigation depends generally upon water from rivers, lakes, and wells, and irrigation systems are expensive; many of the developing countries that have the greatest need for such systems can ill afford them.

It is evident that, as the population expands and the quality of life improves, more water will be needed. Yet the annual amount of rain that falls upon the earth does not vary greatly. Water is exhaustible, and we are not conserving it very well.

Soil too is exhaustible. There is just so much of it; we cannot increase our acreage. When soils are depleted through lack of conservation, it takes a long time to build them up, more time than humanity, with its growing population, can count on, and even those who are most ready to mention substitutes have no suggestion to make in this case. In the United States, with advanced methods of agriculture, including the use of fertilizers, the acreage has been so productive that the government has paid farmers to keep land out of production; today we have a reserve called the "soil bank." Yet this reserve has its limits, and there is no similar reserve in other parts of the world.

The agricultural products of soils and water are a renewable resource, in contrast with the nonrenewable resources we have discussed in earlier chapters. Like the nonrenewable resources, food supplies are becoming increasingly important everywhere. One of the great resources remaining to the United States is the productivity of the Mississippi and California valleys, where, under the application of the technology of American agriculture, the yield year in and year out has been more per acre than any other place in the world. This yield may be a vital factor in the future in this country's trading of what it has for what it needs.

If there are to be more and more people, where are they to get their food? Will the oceans serve as a source of food as well as water? Will this source prove to be inexhaustible?

It is encouraging to note that improved agricultural practices are spreading. The development of short-stemmed and hardy grains, which with moderate use of fertilizers give considerably increased yields, bids fair to assure enough food for the existing population of the world. It is encouraging to see that smelters and other industrial developments are now constructed not to harm the land, that water pollution is decreasing in some areas, and that dams and desalinization plants represent intelligent efforts to insure an adequate water supply.

Even so, the population of the world cannot continue to expand indefinitely without ultimately facing shortages of soil and water. There is no substitute for clean water or for rich soil. Both exist in finite quantities. Both must be cherished by man.

12

Economics and
the Extractive Industries

The common measure of the health of a nation's economy is its Gross National Product, or GNP. The GNP is the value of the services and goods available to a nation.

Services

Many varieties of services are necessary or useful in varying degrees. In general, services may be classified as the performance of the following kinds of work:

• Maintenance. This includes the professional services of doctors and lawyers as well as the manual services of all repairmen such as electricians, plumbers, and mechanics.

• Distributive. Distributive services are provided by the transportation and communication systems and by salesmen, advertising agencies, and all people engaged in marketing.

• Controlling. Controlling services, which may either constrain or encourage the GNP, are primarily performed by government and thus described in Chapter 13.

• Productive. Services which produce may be those of labor or management. They are also performed by people engaged in creative work and by those who in any society, socialistic or free enterprise, invest and risk their capital.

Goods

Without goods there is little need for services. Goods are of two types, *primary* (such as foodstuffs, minerals, and crude petroleum) and *processed* (such as automobiles, refrigerators, clothing, and refined fuels and electricity).

Fig. 12-1. Energy use and Gross National Product are closely related. GNP, in billions of dollars at 1958 value, is shown by the dotted line. Energy consumed, in quadrillions of BTU's, is shown by the solid line. GNP and energy use are those of the United States, beginning in 1920 and estimated to 1990. Source: Hollis M. Dole.

Billions of 1958 dollars   Quadrillions of BTU's

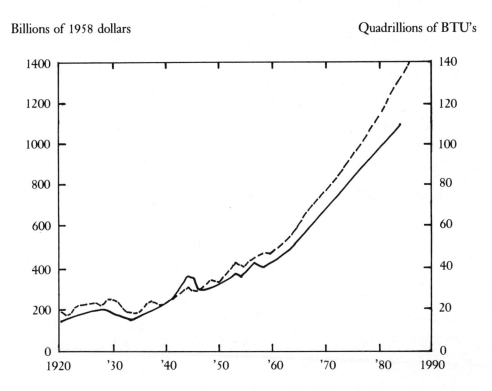

The extractive industries are the originators of practically all important primary goods except food and natural fibers. While the value of the raw materials in processed goods may be, and usually is, only a minor fraction of the sales value, yet without such materials as copper, iron, and mineral pigments, and energy, there can be few processed goods—and no transportation. Thus without the extractive industries there would be scant necessity for services, and a low GNP would result.

Some extractive industries also perform services. Some steel companies refine and sell iron ore wholesale; some petroleum companies transport and sell at wholesale or retail the products they obtain by processing crude oil and natural gas.

Most of us are unaware of the extent to which minerals affect our national economy. How many people ever think of zinc? Yet it is essential for the functioning of the printing industry, so necessary for communication and education, and the tire industry, on which transportation and distribution so greatly depend. Iron, as another example, is required for transportation and by the building industry.

Although zinc, iron, and other raw materials are essential for manufactured goods, their relative cost is small. The steel in an automobile selling for about $3,000 may cost $300 at the steel mill, and this price of $300 includes costs of labor, management, and capital investment, costs incurred not only in the mill but all the way back to the mine, where the iron ore from which the steel was made cost about $60 to extract.

Of the $300 that a person pays for a refrigerator, only about $20 to $25 can be charged to the cost of the steel when it left the mill. The rest, approximately $275, represents payroll (or energy) costs accumulated along the complex route of production and distribution. Without steel, however, there would be no refrigerator or automobile as we know them, and without such processed goods the standard of living would be greatly below that now enjoyed in the developed nations.

GNP–C

A country's GNP divided by the number of its inhabitants is the average per capita gross income of the nation. This average per capita distribution of the GNP (GNP–C) is accepted as a measure of the quality of living of a nation's inhabitants. Of course the GNP is never distributed equally among its inhabitants. The disparity between low and high income is greatest in the case of a country whose economy relies upon a single commodity, such as tin or copper.

In any society, whether it be reactionary, communistic, socialistic, or free enterprise, economic health depends on the following factors:
• An educated and productive labor force that is fully employed.
• A creative and imaginative management staff, concerned with the welfare of workers.
• Available risk capital.

The average quality of living will be highest where there is mutual trust and understanding among labor, management, and those able and willing to invest.

GNP, GNP–C, raw materials, and society: examples of their interrelationships

Russia is probably the world's greatest storehouse of undeveloped resources, and some people believe that it is a society well organized for the production of goods. Yet news coming out of Russia indicates that there are great consumer demands that are not being met. In Russia the cost of services is low—the average worker makes less than $150 a month and a doctor less than $200. Prices are high—a pair of workman's shoes costs almost a week's average income. The GNP–C is $1,200.* Why does the Russian farmer produce only about 10% to 15% as much wheat per acre as the American farmer? Why is Russia, rich in raw materials, unable to satisfy its consumers' needs?

China is probably the second greatest storehouse of undeveloped mineral supplies. It has an abundance of human energy but little mechanized energy. Apparently no one has to be hungry in China today and the nation possesses many millions of dedicated workers. Yet the GNP–C is estimated to be less than $100. What is lacking?

Chile is a nation in transition. It has depended heavily for foreign exchange upon one resource, copper. When the price of copper was high, Chile's GNP grew 2.17% per year—but its population at the same time grew 3.1%. Its GNP–C is only $510. What is needed?

Zaire is rich in minerals but not in energy supplies. Without the money to pay for imported energy, it must depend upon human labor. Why does it have great difficulty in obtaining the risk capital that it requires to develop its resources? What does it need to improve its quality of living, which is low although slowly rising?

Japan has inadequate mineral supplies and must import most of the energy it uses. It has dedicated workers and risk capital that is aggressively utilized by imaginative management. Its economy has had the most notable growth in GNP–C of any nation during recent decades. Why?

The United States has the highest GNP–C in the world today—$4240. This nation has spent a larger part of its GNP on education than any other nation; at least twelve years of education are not only available to every citizen but required by law. A half century ago the United States had vast supplies of basic minerals and energy. It developed mechanized energy that greatly increased the productivity per worker. Today the United States possesses less than a dozen of the hundred raw materials required by its factories and workshops. It must import; in recent years the goods it has imported have cost more than the goods it has exported; it has found itself increasingly hard put to produce at competitive prices in world competition. For example, on the west coast most of the nails available for sale come from Japan, and a roll of barbed wire made in Japan is 15% to 20% cheaper even after being transported as far as the Rocky Mountains than a similar roll made in the United States. Why in fifty years has such a change occurred?

* GNP–C figures are those of 1969, according to the International Bank for Reconstruction and Development.

Fig. 12-2. United States consumption and production of minerals since 1950 and projected to 2000. The deficiency of domestic production is made up by imports which affect adversely the nation's balance of payments. The amount of the deficit is shown for 1950, 1970, 1985, and 2000, in billions of dollars at 1970 value. Source: United States Department of the Interior.

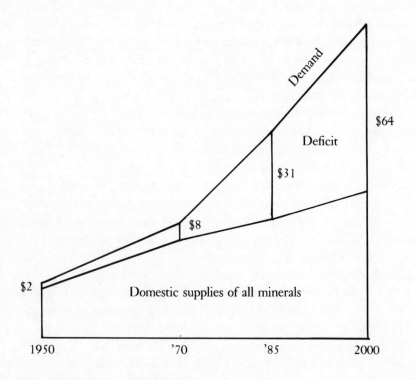

The interrelationships of GNP, GNP–C, and raw materials are close; the ways in which various countries have treated their resources of minerals and energy have produced widely differing results.

Economic developments in the extractive industries are always complex. The opening of a mine or oil field generally starts with an idea. The project then requires risk capital and competent management. The old saying that "it's a poor mine that will not withstand some mismanagement" is no longer true; ore deposits are so low in grade that mismanagement cannot be tolerated. Dedicated workers are also needed, and must receive adequate pay. As wages go up, however, so also must the productivity per man hour if inflation is to be avoided.

Fig. 12-3. Imports in 1970 of minerals vital to the United States. Source: United States Bureau of Mines.

	Imports		*Imports*
Columbium-tantalum . .	100%	Bauxite	87%
Graphite		Cobalt	93%
(natural amorphous) .	100%	Fluorspar	54%
Industrial diamonds . .	100%	Rutile	100%
Quartz crystal	100%	Lead	58%
Tin	100%	Cadmium	21%
Asbestos (long fiber) . .	83%	Zinc	66%
Beryl	95%	Tungsten	52%
Platinum metals . .	99%	Barite	50%
Manganese ore	100%	Mercury	66%
Mica (sheet)	100%	Gypsum	39%
Antimony	91%	Ilmenite	11%
Chromite	100%	Iron ore	34%
Bismuth	93%	Copper	0%
Nickel	90%		

Labor and wages

The largest item in the cost of production of raw materials, except for some petroleum and natural gas, is the cost of labor. So long as productivity per man hour remains high, workers earn the money with which to buy and thus add to the GNP. When productivity goes down, wages and standards of living follow.

Before John L. Lewis became president of the United Mine Workers' Union, coal miners (in the United States as well as in most other countries where coal was mined) worked long hours for low wages under conditions that showed complete disregard for their comfort, safety, and health—coal mining was hard, unhealthy, and dangerous. When Mr. Lewis finished fighting for them, coal miners, especially in the United States, had vastly improved working conditions and higher wages, though there were fewer miners.

The effect on the GNP was both beneficial and deleterious. Remaining jobs were better, and better paid. However, the cost of coal rose to the point where coal was priced out of much of the energy market it had previously enjoyed, which was inherited by other sources of energy, and improved mechanical mining partially replaced the miners. One miner using a machine took the place of several miners using hand tools.

Can anyone say that Mr. Lewis was wrong in improving conditions for the miners because the resulting higher cost of the miners' labor contributed to the pricing of coal out of parts of the energy market and led to the sacrifice of some of the jobs? Can anyone say that management was wrong in attempting to keep labor costs down in an effort to maintain a competitive price during a time when coal companies were trying desperately to retain their markets?

Henry Ford, Sr., although not analogous to Mr. Lewis, was another man who had a long-range effect upon labor and wages. In his automobile plants Mr. Ford raised wages in the early part of this century to the then unheard-of minimum of $5 a day, on the thesis that more workers would have more money to buy more cars.

In countries such as India and China, where human energy is abundant in the form of unskilled labor, industry pays low wages but must employ larger numbers of people than do comparable mines or industries in the developed nations. This means that the relatively low cost of the individual's labor is offset in part by the greater numbers of persons who have to be employed to perform each task; also, the lack of skills and education, and the commonly faulty diet, all reduce efficiency. Especially in the extractive industries manual energy is seldom as inexpensive as other forms of energy because it is not as efficient—productivity per man hour is low. How many men would be required to displace the large shovels and trucks in a strip coal mine or a disseminated copper ore deposit where one hundred thousand tons or more of ore may be extracted every day? Could these large mines be run on their present scale without mechanical equipment and abundant mineral energy?

Where labor is unskilled and wages are low, labor has little money to purchase other than the immediate necessities for living, and the GNP is low, although the cost of many commodities in the underdeveloped nations is not as low as might be supposed.

The ability of Japan to produce goods and materials at prices below those of its competitors is generally explained as possible because of the high quality of its "cheap" labor. It is true that basic salaries in Japan are considerably less than those for comparable work in the other industrialized countries, but the average workman gets fringe benefits that far surpass similar benefits elsewhere. These benefits include subsidized housing, allowances for commuting, rewards for good attendance records, company savings plans offering high interest and loans at low interest, gifts to mark special occasions such as a child's graduation from school, off-duty classes ranging from foreign language study to flower arrangement, a hall for weddings, lounges for rest, sport facilities for recreation, and elaborate vacation retreats by the sea or mountains where bed and two meals are available for $3.50 a day. Such non-cash benefits boost the cost of labor by at least 20% and at times considerably more. Also, frequent large bonuses, often excluded in a discussion of Japanese labor costs, may in some instances amount to as much as 50% of the salary base. Workers in Japan are expected to stay with a company for life, and labor tends to be a fixed charge because employers are not free to discharge workers during slack periods. The system is paternalistic. Job security, group solidarity, and technical innovations are among the objectives stressed (*69*).

However, Japan has had to import everything except rice. Its successful economy has been dependent on the low cost of imported energy and other raw materials. What will happen during the rest of the '70's if costs of imports rise so that the Japanese can no longer export at such competitive prices?

Since the cost of labor enters to such a large degree into the cost of mineral commodities, the tendency exists to blame labor costs for the disastrous effects of inflation. It is true that in many places and industries the cost of labor has had a faster rate of increase than has the output of commodity per man hour of work (*21*), which contributes to inflation and cannot continue for a long period of time without causing the products to be priced out of the market. Is labor really to blame for this situation? Or is the basic cause to be found in government overspending and abundance of money? So long as productivity per unit of labor increases, labor and management and capital should all share the benefits—all three will profit. If productivity per unit of labor does not increase as cost increases, the result may be an industry that cannot compete—and then labor and management and capital will all lose.

An obvious result of high labor costs is the tendency of industry to mechanize wherever possible. Since automation means that numbers of routine tasks in mines, oil fields, and plants can be accomplished in a fraction of the time and at a fraction of the cost required by hand labor, management must seek automation in order to remain competitive. But industry cannot afford to become fully automated overnight, for social upheaval and economic depression could well result. How fast should automation be permitted to proceed? How can older plants remain competitive with newer, more highly automated plants? Automation means the use of mechanical energy for hand labor, and where properly used brings about more leisure time for labor. Should labor encourage automation, with its higher pay for jobs with greater skill requirements, in order to improve the standard of living? Can labor afford to maintain the status quo? Can labor afford to increase the cost of goods and services by such practices as "featherbedding"? *

In most western countries, mine, oil field, and refinery workers are organized into local labor unions that enjoy considerable autonomy and control over their own affairs. In some countries, however, labor unions in the minerals industries have combined into strong national organizations whose leaders thus

* "Featherbedding" is the maintenance of unneeded jobs by means of which the employer is forced to pay for services that are not necessary or are not performed.

have it within their power to dictate the terms of labor settlements and to strengthen or weaken the national economy. Labor as organized in the United States is powerful. Labor leaders should keep in mind that the safety, welfare, and income of workers depend upon productivity per unit of energy expended. If productivity decreases, basic industries go elsewhere and jobs are lost. Labor should also remember that it has billions of dollars of reserve and pension funds invested in American corporations; it has risk capital at stake, and the welfare of labor as well as the GNP is adversely affected if labor asks more than the demands for its products warrant.

When large national unions decide to strike, they are able to paralyze the mineral industries and thus influence entire segments of the national economy. This great concentration of power in the hands of a few labor leaders is a matter of concern to many citizens in the United States. The average citizen does not ordinarily condone work stoppages, especially those of jurisdictional types; he knows that strikes are wasteful and disruptive, and that he pays for them in both currency and inconvenience. He may justifiably question the necessity for many strikes. For example, is it necessary that our copper or steel industry be immobilized for extended periods every two or three years during labor negotiations? Is labor so grossly underpaid and mistreated? He may wonder why powerful labor unions should not be subjected, for the national good, to anti-trust laws or other controls.

The main purposes of a union are to stabilize employment, to improve wages and working conditions, to assure continuing benefits in case of illness or economic depression, and to obtain adequate retirement pay. In some instances, however, unions have insisted upon "featherbedding," or participation in the formulation of policy decisions (which is normally a function of management), or profit sharing, or shorter and shorter working hours. Shorter hours are demanded in order to spread available jobs, which shrink in number as automation achieves its aim of fewer man hours per unit of product. Shorter hours in many cases have resulted in the practice of "moonlighting," the holding of two or more jobs simultaneously, which defeats the purpose of shorter working hours by cutting down on the number of jobs available to the unemployed.

When it is forced to grant shorter working hours, thus increasing costs, management in turn is stimulated to seek further labor-saving practices. Consequently the process of automation is accelerated and more jobs are abolished. The result is a vicious circle that causes unnecessary difficulties for both labor and management, adds to costs, and creates inconvenience to the consuming but never consulted public.

The high cost of labor in the United States unquestionably has had a direct bearing on the production of domestic minerals and is one of the reasons why many companies have turned to foreign operations. However, in recent years political unrest and well publicized examples of expropriation of mineral properties abroad have dampened enthusiasm for foreign exploration and mine development. Only in nations such as Canada, Australia, Ireland, and South Africa, where the governments have long histories of stability in their treatment of foreign investments, is development in the extractive industries continuing at a rapid pace.

Although steel in the past few years has been imported into the United States from as many as fifty different nations, the great bulk of imports was from Japan, West Germany, and Luxembourg. These are heavily industrialized nations with labor forces as efficient as any in the world; base pay is, however, somewhat lower than in the United States, although in Japan fringe benefits tend to equalize the labor costs. The graph in the figure, reproduced from *Steel Facts* (2), shows the basic labor salaries in the United States in relation to its two leading competitors, indicating how these two nations, even with large fringe benefits, have been able to undersell the United States. Although the United States may build some of the best and most modern and efficient plants in the world, they face competition that is formidable, especially when it is aided by inflation at home. Can the United States maintain a viable and healthy iron and steel industry and pay the higher wages in any way other than by keeping man-hour production higher than that of the competition?

The energy of labor represents cost to the management, to the purchaser, and to the consumer, and it represents wages to the workman, who is also a consumer. Speakers for labor unions correctly point out that prosperity depends upon the continuing purchasing power of the workingman, and that the high standard of living in the industrialized countries results to a great extent from progressive, militant labor policies that have given workingmen the money to purchase and the leisure time to enjoy the products of industry. Can this remain true when productivity per man hour of work does not keep abreast of earnings? If labor is responsible for the increase of costs to levels where the products are noncompetitive, can this policy be anything except self-defeating? Can a nation like the United States afford to let industries that supply basic raw materials, such as iron and steel and base metals, move abroad? These industries, with their management, technological skill, and capital, would be welcomed by many nations, including some that are underdeveloped.

All management is not corrupt, grasping, and anti-labor; all union leaders are not corrupt, power-mad, and anti-industry. Many men, like John L. Lewis and Henry Ford, Sr., have endeavored to improve the life of the workingman in the belief that better working conditions and earnings mean better business.

In an enlightened nation, intelligent workers must share in the general prosperity of the country. They demand and deserve safety at work, security in

Fig. 12-4. Hourly employment costs in the steel industries of Japan, West Germany, and the United States, from 1960 through 1970. Source: *Steel Facts,* summer 1971, American Iron and Steel Institute.

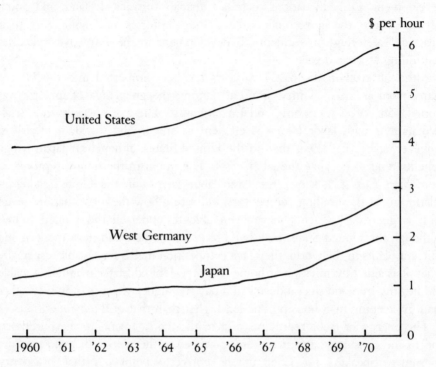

the form of fringe benefits, and good living. But if the cost of labor becomes excessive through inflation or any other cause, a nation cannot produce in competition with other countries. The management of its industries will then be forced to move toward automation or try to produce elsewhere under more favorable conditions. And strikes, though at times they may be necessary, are at other times unnecessary, and are at all times disrupting to a nation and weakening to its economy.

In order to assure United States industry a competitive position both domestically and internationally, to assure labor equitable treatment, to assure consumers a steady supply of materials and energy, and to insure investors a reasonable return on their money, it is essential that labor and management cooperate closely in mutual confidence and trust. What alternatives are there to such voluntary cooperation but government controls, communism, or dictatorship?

International labor. International labor organizations have made many efforts to improve the lot of the workingman, especially in the poorer nations of the world. Among these organizations is the International Labor Organization (ILO), which came into being in 1919 after the First World War. It was founded upon the concept of lasting peace based upon social justice. Its primary efforts have been devoted to eliminating conditions of poverty and to improving the general welfare of workers.

In 1946 ILO became the first specialized agency affiliated with the United Nations. In part it owes its success to its unique combination of members from governments, employers, and workers. Its principal offices are in Switzerland and most of its meetings are held there, although some are held in other places throughout the world.

ILO attempts to improve manpower organizations by providing education in vocational training, productivity, management development, cooperation, handicrafts, social security, labor conditions, and administration. It has been meeting head-on the paradox that has continued to plague the developing world—that of an abundance of laborers on the one hand and, on the other, a shortage of jobs and trained people at all levels. In a noteworthy attempt to develop needed skills in industry, ILO in 1965 formed an International Center for Advanced and Vocational Training at Turin, Italy. The ILO has not, however, come to grips with the real problem of population control.

Another international labor group that is of special interest in the United States and that directly impinges upon the mineral industries is ORIT, the Interamerican Regional Organization of the International Confederation of Free Trade Unions. This organization, with offices in Mexico City, is concerned primarily with the establishment of free trade unions in Latin America. ORIT has steadily, and at times militantly, opposed all attempts to exploit labor politically by either the left or the right, and is constantly making efforts to improve conditions for workers throughout its sphere of influence. Particularly commendable have been its endeavors to free Cuban labor from the yoke of the Castro brand of communism.

ORIT is in an especially favorable spot to better the circumstances of labor in Latin America because it is able to speak as one workingman to another, thus avoiding the implications of paternalism by government, by big industry, or by charity under any name.

A somewhat different approach to international labor problems has been slowly emerging in the United States. As companies establish branches abroad in order to take advantage of better labor conditions, larger markets, and lower costs, the number of jobs available is reduced in the United States while at the same time it is increased abroad. As many companies in the extractive industries have become international or even multinational in scope, domestic labor unions are able to exert less and less influence over the total activities and policies of the companies. Some labor leaders do not like this trend and are now considering the possibilities of forming international or multinational unions to cover the activities of single companies or even of individual industries.

Another recent development is the migration of labor from one country to another. West Germany, in the European Common Market, expanded its industry so rapidly that shortages of both skilled and unskilled labor resulted. Unskilled workmen were brought in from Yugoslavia, Greece, or other countries where work was scarce. After the work term expires, the laborers return to their own lands, generally with considerably more money than they could have earned at home. Another example of migration of labor from an underprivileged nation to an industrialized one is the movement of Yugoslavian workmen to Israel, where laborers are badly needed. Numbers of skilled workmen have also migrated to West Germany from the somewhat depressed industrial areas of Great Britain, though many became dissatisfied and soon returned home. Migrant labor from Mexico has long been used in the United States, but here in recent years the practice of importing laborers has been heavily attacked by labor unions.

The policy of importing workers from depressed areas to places where labor is needed is becoming an established method of employment. It is not without problems, but it also has advantages.

Investment and capital

A need for money exists in all parts of the minerals industries. To develop a mine or an oil field may require only a few million dollars under the best of conditions, but it may require many millions. To bring the North Slope oil fields of Alaska into production will require a minimum of three billion dollars for pipeline construction alone. When the costs of drilling, storage, port facilities, housing, and supply are added, the total will be staggering. Most mining enterprises are small by comparison, although the development of a large open pit mine may cost one hundred million dollars or more.

The large sums of money necessary to develop an oil field or a mine can come from two sources only, private investors and government agencies. In the industrialized nations of the west much of the capital is obtained from private sources, supplemented by government loans and at times by government insurance against loss of capital because of expropriation abroad. In socialist and communist countries, governments and government loans supply the capital. In underdeveloped countries where needs for capital for all purposes far exceed the money available, funds are commonly borrowed from one of the large international development banks, from other governments, and, as far as possible, from private sources (24).

Fig. 12-5. Direct investment of United States companies in petroleum industries of other areas in 1968, in millions of dollars. Source: United States Bureau of Mines.

Canada	4,088
Latin America	2,976
Other Western Hemisphere	667
Europe	4,640
Africa	1,567
Middle East	1,654
Far East	1,146
Oceania	646
Trading and shipping companies	1,451
Total	$18,835,000,000

Fig. 12-6. Direct investment of United States companies in mining and smelting industries of other areas, in millions of dollars. Note the increase of 58% from 1964 to 1969. Source: United States Bureau of Mines.

Area	*1964*	*1965*	*1968*	*1969*
Canada	1,671	1,755	2,636	3,315
Latin America	1,098	1,114	1,402	1,346
Europe	56	55	61	
Africa	356	361	387	427
Australia	100	161	365	395
All other countries	265	298	519	
Total	3,564	3,794	5,370	5,635

The money invested in new mining and energy projects, with the hope of repayment from profits but subject to the hazards of ownership, is called risk or venture capital. No country can grow economically without having risk capital available, for without it there would be no large mines or plants and few discoveries of new ore bodies. Nothing ventured in capital would bring nothing gained in the establishment of business and the finding of valuable mineral commodities. It is also a fact of history that when a government supplies the capital, those in charge of the responsible bureau prefer the politically safer practice of spending the capital in wages rather than on the risky uncertainties of exploration. It has been individuals and private corporations operating in a free economy that have borne the greatest risks of capital investments in exploration and development (34).

The need for risk capital is very great. For instance, geothermal power at the California Geysers area reached the capacity of 400,000 kilowatts in 1974, but the first experimental well was driven in 1955. Risk capital developed the project for years when there was no return from it and no certainty of its eventual success.

The need for risk capital is especially acute in underdeveloped countries that are trying hard to industrialize and create jobs for their unemployed (*16*). Most underdeveloped countries are anxious to attract foreign capital to help in the development of their natural resources. To attract this capital, many plans have been advanced, generally including land guarantees and periods of exemption from taxes and duties. A few underdeveloped nations refuse to permit foreign capital to invest in their mineral developments for a variety of reasons, including national pride, fear of exploitation by foreigners, security of supply, diverse political ideologies, and major social and political problems.

Since no one can foretell with certainty the extent of a buried pool of petroleum or of a buried body of ore, the term "risk capital" for investment in the extractive industries seems particularly appropriate. Two recent examples illustrate further uncertainties that must be faced by those who invest risk capital in the minerals industries. When the state government of Alaska held its auction of oil leases, the competing petroleum companies had to submit bids under a number of handicaps. Before the bids were opened, none of the competitors knew which of them would win which lease. Each who bid on a parcel had to have cash in Alaska ready to pay the state in case his bid was accepted. The winning bids totaled almost one billion dollars. How much risk capital was temporarily deposited in local banks? How much interest was lost by the unsuccessful bidders? The winning bidders, in addition to the prices they paid for the leases, will have to pay royalties on any petroleum found and extracted. All those who bid had already spent considerable sums in geological exploration of the area.

More recently the federal government auctioned oil leases off the Louisiana coast in the Gulf of Mexico. Here again winning bidders paid approximately one and a half billion dollars for the rights, with royalties to follow if they were successful in extraction.

Because of these high risks, those who invest in the extractive industries expect a fairly high return after taxes, especially since they face not only the uncertainties of exploration and political difficulties but also the fact that even the best of ore bodies is eventually exhausted. Where danger of expropriation exists, the investors plan to recover their capital and interest in no more than two or three years.

The typical successful mining enterprise generates gross annual earnings of about 25% of the invested capital. Of the 25% gross profit, probably about 50%

must be set aside for income taxes. About half of the remainder ($12\frac{1}{2}\%$ net profit) is reinvested in the business, going into reserves for exploration, the purchase of new ore bodies, or capital investments in modern equipment. The other half of the remaining profit, or about 5% to 7% of the investment, is distributed as dividends on which further taxes are paid.

In recent years the danger of loss and the degree of risk involved in mining have been somewhat reduced for the average investor, as have also the possibilities for spectacular gains; the result is that small mining and oil companies are now far fewer. Mining and petroleum companies have followed the pattern of other industries; they have become so large that they are able to spread their risk over many areas and among different commodities. This adds security against a property loss that would be fatal to a smaller company. For instance, Anaconda Mining Company, because of its size and diversity of investments, was able to survive the expropriation of its Chilean properties, although these constituted something like 60% of its total capital assets. Likewise, the stock of the much larger British Petroleum Company hardly fluctuated even after the company-owned oil fields in Libya and Iraq were expropriated.

Considering the degree of risk, it is reasonable to expect that those who have the capital will search for the industry and the area that will provide the best opportunity for return on the investment. If a country limits the net profit after taxes, or prevents free movement of capital, then the investment will be made elsewhere. Competition for venture capital always exists and the capital will go where it is most welcome.

After capital for investment in the extractive industries is obtained it may be spent in three ways:

• For exploration. The average cost of discovering a mine worth developing is now close to twenty million dollars.

• For the construction of mine and plant, including pipelines for the collection and transmission of oil and natural gas.

• For investment in inventory and operating expenses until the product furnishes sufficient income to cover the cost of management, labor, and maintenance.

Because all industrialized nations have been increasing their consumption of raw materials in recent years, they have also had to increase their investments abroad. These investments are made primarily in places where economic and political climates are stable and favorable. Risk capital stays away from areas where expropriation threatens. For example, until about 1950 Latin America was the preferred investing ground for North American capital. Since then, as a result of political unrest, expropriations in Cuba, Bolivia, and Chile, and indications of expropriation in many other parts of Latin America, venture capital has become wary and is seeking more receptive areas such as Canada and Australia. Comparisons of the totals invested in major areas of the world between the years 1960 and 1967 show that Latin America has the smallest increase of all, only about 20%. Investments in Latin America in the mining and petroleum fields continue to decline. Only in Brazil and Mexico is risk capital still available for the extractive industries.

Profits. Without profits there could be no capital to invest. Profits can be defined in several ways; financially, profit is the money remaining after costs of operation, including taxes. Thus for an individual or a society it is the margin earned over what is needed merely to exist. Any denigration of profit ignores the fact that, regardless of the kind of economic system under which they live, people anywhere must have profit if they are to exist above the subsistence level.

Responsibilities of companies. Corporations were designed for earning money; they were not established to implement social reform or as charitable institutions. However, changing conditions have forced most companies to recognize the fact that they must fit their pattern of organization and operation to national aspirations. Their executives are willing to discuss methods of bringing this about, and most company budgets reflect the modern trend by allocating funds to charity and social improvement.

People everywhere for generations have tended to regard minerals and natural resources as property of the state rather than of the individual. The result has been that companies engaged in the extractive industries have been unable to act as freely as some of them would like. The companies thus have developed a pattern of operations that is generally applicable everywhere.

Drolet (*14*), in discussing mining companies in Canada, lists ten activities that he believes should be carried on by any company operating in that nation. These points* have been considered many times in the industry. Their implementation, especially by companies operating in foreign countries, would ease many tensions.

• Ensure that local people are associated with management, including local representation on the Board of Directors.

• Allow decision to expand or curtail local operations to be determined locally to a large extent.

• Establish an export policy truly prepared to seize market opportunities wherever possible and not just to supply a parent company.

• Seek assurance of long-term reserves to improve the stability of operations,

* Copyright by American Mining Congress and published by permission.

related communities, and overall activities in the country. With reference to local needs for minerals and metals, firms should try to ensure that supplies are available at reasonable prices.

• Allow local personnel to invest in common shares of the local operation. Foreign financing should include some percentage of local capital. Care must be taken to disclose not only information on the financial character of the firm's business, but also certain operational characteristics such as reserves, output, markets, and plans for expansion or shutdown.

• Concern themselves with the maintenance and improvement of environmental quality, prevention of air and water pollution, and preservation of natural beauty and wildlife.

• Participate actively in local community affairs with a sensitivity to the aspirations of the local people.

• Carry out research and development locally and allocate sufficient funds to solve problems unique to local operations. Do not just apply foreign solutions indiscriminately.

• Cooperate with local universities by sponsoring research projects and providing visiting lecturers.

• Use the official language of the host country in so far as possible.

Conclusions

A nation must have minerals and low-cost energy available to it in order to have a high GNP and a high GNP–C. It must therefore either have its own supplies or import. If it has its own supplies, it must encourage risk capital to find and develop them. If it imports, it must also export in order to maintain a favorable balance of trade, balance of trade being exports minus imports. In either case, it must have intelligent management and well paid, productive labor.

Since we are here concerned only with the relationship of economics and the extractive industries, we are not discussing general economic theory. We are, however, rejecting one modern concept of economics—that of an ever expanding economy, with plenty for all. That concept is based upon the availability to everyone of unlimited supplies of energy and other nonrenewable raw materials, and the concept is false because the finite earth's supplies of raw materials (minerals and energy) are limited.

We accept the classic law that supply and demand need to be in approximate balance. All people want more things from the earth, and continued population growth increases the demand on the earth's supplies. As demand increases, fewer materials are available to the individual and prices to the consumer rise. The standard of living goes down. If the supply of a raw material is insufficient to meet a nation's demands, then the price will go up until the supply and demand are in balance; that may mean that the cost will be so high that the poorer segment of the population will have to do without the commodity. The GNP–C will fall.

Nations take various actions to supply their citizens' demands, either by development of national resources or by importation. They take actions that affect the productivity of their workers, the control of management, the investment of risk capital, and the cooperation of labor, management, and investors. All of these actions are political. The economic welfare of a society is largely dependent upon its politics.

13

Politics and
the Extractive Industries I

〜〜〜〜〜〜〜〜〜〜〜〜〜〜〜〜〜〜〜

A society's political actions, upon which its economic welfare depends, are the ways in which it conducts its government on three levels—local, national, and international. Political actions are taken by officials in power—by legislators, judges, and especially chief executives and the bureau administrators appointed by the executives. Those in power are motivated by many often conflicting factors: the strengths and weaknesses or needs or demands not only of the society as a whole but of different groups within the society. For this reason, political actions are usually compromises. Political actions are also likely to be whatever those in power believe will keep them in power.

In a society in which the electorate chooses its representatives, political action, although carried out by the officials in power, is determined by those who put them there—the voters, informed and uninformed or misinformed. The degree of education of the electorate is therefore the key to the quality of the society's political actions. Uneducated voters cannot vote wisely. Only if his constituency is relatively well informed can we expect an elected official (who may also have a voice in making bureaucratic appointments) to act for the economic welfare of the society as a whole rather than to work for the advantage of a particular group in order to secure its votes. Only an enlightened electorate can realize that its welfare depends on its voting, not primarily for the self-interest of local segments, but for the benefit of the GNP and the economics of the whole society.

The United States has had until recently supplies of raw materials that have made us largely independent of foreign imports. At the same time we have had productive workers, effective management, and imaginative risk capitalists. These fortunate circumstances have provided the nation with the highest GNP–C in the world.

Either as cause or as effect (or both) of all these factors, we have spent a higher percentage of our GNP on education than has any other society in history. An article of political faith among American voters has been liberal support of the educational system. We might therefore expect that the United States, with a system providing widely available education, would be able to take relatively wise political actions. But the educational system runs the gamut from excellent to poor. Many voters are less than adequately informed. Our political leaders, whose every act is directed primarily toward re-election, reflect the electorate's wide variation in education. Our political actions are both wise and foolish, good and bad.

Good political action benefits all three factions of the producers of goods, because labor, management, and capital are interdependent and all necessary to a healthy GNP, without which everyone in the nation suffers to some degree. Bad political action discourages the production of goods, especially the basic goods of the extractive industries.

In this chapter and the next we discuss the results of political actions, especially in the United States, that have a strong influence, good or bad, on the extractive industries: the percentage depletion allowance; taxes, including duties and tariffs; quotas; subsidies; the balance of payments; inflation; stabilization and pricing; and stockpiling.

Percentage depletion allowance

The percentage depletion allowance is an almost perfect example of how an informed electorate can benefit by encouraging its extractive industries. Yet throughout 1972 there was in the United States much political activity brought on by the widespread cry for its abolition. This cry is still heard from those who are ignorant of what the percentage depletion allowance is, why it was instituted in the first place, and what the effect (on the extractive industries and on the nation) would be if it were canceled.

Percentage depletion allowances relate only to the extractive industries. In those industries, such as iron and petroleum, where there are both extractive and other operations, the percentage depletion allowance applies solely to the extractive part of the business.

In other industries, the law permits, or even insists, that whenever factories and stores and machinery and equipment are paid for in risk capital, these initial costs be capitalized and then amortized over a period of time as an expense against gross income. The purposes of this depreciation charge are two: to encourage venture capital to build a business and to insure that the business has reserves for the replacement of old equipment and buildings when they are no longer economic to operate. Without risk capital the business could never have started; without new equipment the business would in time cease, and its jobs and products would be lost to the GNP.

While this law is reasonable, it caused in the extractive industries an extremely complex structure of accounting, legal work, and governmental bureaucratic supervision. Its application to the extractive industries had to be based on estimates of recoverable oil and gas reserves, or of ore from a deposit, made property by property. By the 1920's, when the amortization of depreciation system had become established in the United States and its impact on the costs of running a business had become noticeable, depreciation charges in the extractive industries were cumbersome, unwieldy, and almost impossible to administer.

Through the previous decade, attempts had been made to interpret depreciation charges for the extractive industries in a different way and therefore to lessen the costs of accountants, lawyers, and government bureaus. This action recognized the principle that determination of the values of deposits is impossible, a principle that later became the basis of the percentage depletion allowance. Thus the percentage depletion allowance was originally designed as a substitute for the system of depreciation charges, a system that will not work in the extractive industries as it works in other businesses.

Since its enactment, the percentage depletion allowance has been a far simpler and more effective means of permitting amortization of investments in exploration and development for energy and minerals. It permits the deduction of a percentage, specified by law, of the gross income from a mineral property before arriving at the amount of taxable income. The percentage allowed now varies from 5% to 22% according to the mineral concerned. Under existing laws the deduction is limited to a maximum of 50% of the profit before operating expenses and taxes, and is computed without other allowances for the inevitable exhaustion of the asset. When this deduction was provided, 27½% for gas and oil in 1926 and lesser percentages for coal, sulfur, and metals in 1932, there was a close relationship between the amount established by percentage depletion and the amount allowed in prior years for depreciation.

Before describing how the percentage depletion allowance works, we shall consider reasons first against and then for it.

Many voters and politicians consider the percentage depletion allowance to be the number one tax loophole in the United States. They advocate its abolition citing the fact that, especially in the oil industry, this depletion allowance, being merely a percentage of income from production without regard to past investments, has often had a very uneven effect, sometimes providing large benefits to individuals (such as royalty owners and speculators) who have little or no investment or interest in the industry, and sometimes providing to an investor returns far greater than his investment.

Fig. 13-1. Depletion allowances as of January, 1972, for United States extractive industries. Note that in general more favorable allowance has been given for domestic production. Source: United States Bureau of Mines.

Commodity	% on domestic	% on foreign	Commodity	% on domestic	% on foreign
Antimony	22	14	Nickel	22	14
Asbestos	22	10	Petroleum	22	22
Bauxite	22	14	Phosphate rock	14	14
Boron	14	14	Platinum		
Chromium	22	14	group metals	22	14
Coal	10	10	Potash	14	14
Cobalt	22	14	Pumice	5	5
Copper	15	14	Rutile	22	14
Fluorspar	22	14	Salt	10	10
Gold	15	14	Sand or gravel	5	14
Graphite	22	14	Silver	15	14
Iron ore	15	14	Sulfur	22	22
Lead	22	14	Tin	22	14
Manganese	22	14	Tungsten	22	14
Mercury	22	14	Uranium	22	22
Mica	22	14	Vanadium	22	14
Molybdenum	22	14	Zinc	22	14
Natural gas	22	22	Zirconium	22	14

Those who favor it believe the fact that a few have profited enormously to be outweighed by the advantages of the depletion allowance. Tax liability is lowered to about that of long-term capital gains treatment. This encourages the smaller operators, when they make a discovery, to keep their property rather than to sell in order to make a quick profit, and so encourages competition. The depletion allowance is especially helpful to wildcatters and small independent oil companies, who benefit far more from it than do large oil corporations. In addition, venture capital is encouraged to explore widely and develop new properties, businesses, and jobs. This was not the primary purpose of the allowance, but it is a byproduct whose value greatly exceeds, for the GNP as a whole, the detriment of disproportionate benefit to a few individuals.

The oil industry furnishes good examples for study and discussion of the percentage depletion allowance. As we have noted (Chapter 8), in general the oil industry is highly competitive, as shown by the gasoline wars frequently fought up to 1973 in various parts of the country and by the bidding for petroleum leases, such as for drilling rights in Alaska and more recently in the Gulf of Mexico. The idea that all petroleum companies act in unison to the public's disadvantage is held by many who do not know that competition for discovery, for marketing, and for leases is keen among the companies. Unfortunately this idea is supported by some politicians who realize that the public has more votes than the managements of oil companies and who do not mind misleading their constituents. It is true that the American Petroleum Institute lobbies, but so do labor unions and other organizations lobby for the benefit of those they represent.

The depletion allowance has shifted much of the oil industry's profit from the manufacturing and marketing phases of the business into exploration and production. However, in the United States the industry as a whole earned only 10.2% on its invested capital in the period 1955–1969, as compared with an average of 11.1% for all American manufacturing industry (K. H. Crandall, personal communication, 1971). If the oil industry earned much less it would no longer attract the capital needed to carry on its necessarily ever expanding search and development.

The problems inherent in obtaining sufficient earnings to justify investments are well illustrated by crude oil and gasoline price increases forced on the industry in 1970 by increased taxes, which in turn resulted principally from the reduction in the depletion allowance from $27\frac{1}{2}$% to 22%. The increased income just about balanced the tax increases. Without it the industry would have become less attractive to capital, exploration would have been set back, and the energy crisis could be even more acute than it is at present.

In the extractive industries there are few comparatively large factories, expensive machine tools, or costly retail outlets to amortize. Instead, there is in the ground an ore body or an oil pool of indeterminate size and value. Exploratory costs have been high; the energy chapters point out that a billion dollars a year go into dry holes in the search for oil and gas. Sizeable sums go into exploration for and then development of each of the other basic commodities. To permit recapture of this risk capital, the percentage depletion allowance is an efficient alternative for the technique of amortization used in other businesses. An additional reason for a method different from that of other industries is the fact that a raw materials industry is steadily consuming its capital, since, as we explained in discussing the extractive industries, there is no turn-in value in an abandoned mine shaft or a dry hole.

What will happen if, as many people advocate, the percentage depletion allowance is removed? The extractive industries will be forced to return to the cost depreciation basis of operation. Accurate estimates of the amount of oil or gas in a pool or of ore in a deposit have grown no easier to make. Accounting, legal, and governmental supervisory work has become no more productive and no less expensive. One thing is certain: any additional costs imposed upon the extractive industries will of necessity be passed along to the ultimate consumer.

How the percentage depletion allowance works. The depletion allowance is a percentage of the gross income and is deducted from the gross profit before computing taxes for the Internal Revenue Service. By contrast, corporate income in other industries is figured on the gross income less amortization of invested capital costs. For example, in an extractive industry where the gross income is one million dollars and the percentage depletion is 20%:

Gross income	$1,000,000	
Operating expenses	400,000	
Gross profit before depletion and taxes	600,000	
Depletion allowance—20% of gross income	200,000	
Taxable income		400,000
Federal income tax (only) at 40%		160,000
Net profit before other taxes		$240,000

In this example, the depletion allowance is less than 50% of the gross profit before taxes, and may be taken as shown.

By contrast, consider an operation where expenses are considerably higher. Here, although the depletion allowed for the commodity may be 20% of gross income, that amount is in excess of 50% of gross profit before taxes, so the full 20% is not allowed.

Gross income	$1,000,000	
Operating expenses	800,000	
Gross profit before depletion and taxes	200,000	
Depletion allowance (20% of gross income, but may not exceed 50% of gross profit)	100,000	
Taxable income		100,000
Federal income tax (only) at 40%		40,000
Net profit before other taxes		$ 60,000

Note that, in the above cases, other taxes further reduce the net profit. Also significant is the fact that, if the relatively simple depletion allowance were canceled, operating expenses would increase (because legal and accounting staffs would grow) without any increase in production.

Significance of the percentage depletion allowance. A review of the effects of the depletion allowance shows clearly what political action may do to encourage or discourage the development of a nation's economy. Venture capital may be kept at home or driven abroad. Within any area where it exists, a mineral may be produced or left in the ground. Businesses may be established or terminated. Workers' income will always be where the political action encourages risk capital to undertake exploration and development and where industry is fostered rather than hindered by taxation.

Taxes

Of all political actions, taxation may have the most widely varying influences. The power to tax is the power to destroy as well as the power to encourage a nation to build its GNP intelligently. Taxes are necessary. Without them we would have neither an educational system nor educated voters, neither public services nor supervision. Yet they can be detrimental to the economy; a local area that imposes too high taxes may drive an extractive industry to move its payroll elsewhere or abandon operations altogether. Taxes can be so many and so complex as to require "services"—tax lawyers, vast accounting set-ups, and complicated procedures in both businesses and government bureaus—that add little to the GNP or to the quality of living although they increase costs to the consumer. Anybody in the United States who has made out his income tax using the simplest of our federal forms may estimate the amount of time, energy, and paper expended by the companies and citizens of the nation each year in this unproductive work. It is additionally depressing to add the amount of work done by all government bureaus, including the Internal Revenue Service.

A tax is the payment of money for use by government for the benefit of the public. Taxes that are commonly assessed against industries, including the extractive industries, fall into four categories:

- Taxes on profits.
- Taxes on property.
- Taxes on sales.
- Taxes on international trade.

Any one tax may be applied by more than one governmental agency, federal or local, and it may be that an industry trading internationally will be under the imposition of taxes by more than one nation.

Taxes on profits are income taxes and are thus related to production.

Taxes on property include the right to carry on a business, whether or not the business is productive of income. There are property, severance, occupation, ad valorem, land use, mine use, royalty, and franchise taxes. The franchise tax is imposed merely for the right to operate; so are the others but they differ from the franchise tax in that (except for royalty taxes) they are based on estimated values of property and of ore reserves. Severance taxes are levied against units of value of an extracted mineral and depend upon the rate of mining and the grade of ore mined or metal shipped. The severance tax is imposed on the privilege of extracting a certain amount of a resource; it has the effect of raising the cut-off grade at which an ore body is no longer profitable to work, but it does not penalize the operation when extraction is curtailed (36). The occupation tax is based upon the value of the ore at the mine (71). The ad valorem tax is levied on the basis of plant value and ore reserves, whether or not the mine is operating, and must be considered an addition to the fixed costs of the operation; it therefore discourages the development of reserves and encourages quick extraction. Royalty taxes are levied on royalties paid by the operator to the owner of the land's mineral rights.

Taxes on sales are collected from the purchaser by the seller, who delivers them to the taxing agency. There are in addition excise taxes on the sales of certain commodities, including at present gasoline in most of the United States.

Taxes on international trade are principally to restrain foreign competition and protect local industry, such as import taxes and duties on imports, although export taxes and duties on exports exist primarily to gain income for the government.

Duties and tariffs. Duties are levies imposed on the import or export of goods into or out of a nation. Tariffs are official lists of duties and for any country show the commodities on which it collects money when they are either brought into the country or sent out. The terms *duty* and *tariff* are often used interchangeably.

Duties are not only taxes on international trade but, when applied to imports, constitute taxes on local citizens (as well as protection for local industries) because the consumer pays more for the taxed product. When applied to exported raw materials they are opposed by the heavily industrialized nations such as Japan and Great Britain that depend upon imports of raw materials and energy.

A good example of the effect of duties is furnished by the German coal industry. Traditionally, German coal has been a basic resource which gave strength to the nation's economy. In the 1950's the Germans found their coal industry, characterized by deep mines and coal of a quality lower than that in parts of the United States, to be at a disadvantage. The response of the government was to keep the politically strong coal miners in the mines, despite cost implications, and, by the imposition of duties on imported coal, to require German steel makers and other coal-dependent producers to use German coal. The result has been the passing on of the basic problems of the German coal industry to its consumer industries, the most important of which is steel. The impact has been pronounced; German steel was losing money by the end of the 1960's, in large part because of the high cost of the coal the industry was forced to use.

The effects of taxes. Another way to classify taxes is to consider who levies them. The three principal taxing agencies are the federal governments, state or provincial governments, and county, city, and local agencies (*47*). All of these may collect any or all of the taxes listed above, but generally certain fields of taxation are held to be the prerogative of certain agencies. For example, the United States government ordinarily confines its taxing program to income taxes, duties, and indirect taxes. The states and local agencies assess taxes on property and the right to use, income taxes, and sales taxes.

The complexity and costs of this unproductive work of tax accounting and reporting are evident, and so is the vulnerability of an extractive industry to unwise taxation policy. If taxation raises the cost of production to a point which prohibits profit, a manufacturing industry may move to another state or country, but an ore body is immovable and can be extracted only at its location. Yet a government is also vulnerable; if it taxes an extractive industry excessively, it may find the ore body abandoned, the equipment, management, and workers dispersed, and all income from taxes gone.

The difficulties encountered by taxing agencies, as well as by companies that must comply with tax laws, are well illustrated by the State of Arizona, which among other taxes levies a straightforward income tax upon profits from mining operations, but permits the deduction of federal income tax payments (*30*). This seems reasonable until the complexity of the federal income tax is taken into consideration; at times the calculations of the amount of the Arizona tax are nearly impossible to make. The items that are especially confusing include loss carry-forwards and carry-backs, prior-year development costs, investment tax credits and how they can be carried forward and back, foreign tax credits, various methods of calculating cost write-offs, mergers, and even percentage depletion when federal income tax paid on the net operating profit of a contiguous mine is required to be deducted before the depletion allowance can be determined.

Michigan furnishes an example of what is happening to some of the mineral-producing areas of the United States. In 1967 Michigan had 12 operating iron mines that shipped 14,033,891 tons of ore, employed 3,430 people, and paid $1,572,550.05 in state taxes. In 1971 state taxes increased to $1,939,530.60; but the number of active mines was reduced to 7, the number of people employed dropped to 2,425, the tons of ore shipped were 12,164,964 (*28*). Excessive taxation encourages foreign ore purchases and curtails available jobs; it affects adversely the nation's balance of payments.

An example of the direct effect of taxation on mineral production is seen in the State of Minnesota, where for many years the ad valorem tax has been applied to reserves of iron ore. This somewhat unusual form of taxation was established under the "Natural Heritage Theory" and has been easy both to apply and to collect. It was originally designed to encourage quick extraction of ore and to obtain maximum revenue for local communities as well as for the state, and proved to be an effective way of stimulating ore removal during the early days of mine developments. But the time for such stimulation is long past and the tax now has the deadening effect of preventing exploration and maintenance of the state's direct shipping ore reserves. The direct shipping ores to which the tax is still applied are no longer in demand; they are difficult if not impossible to sell, and the profit on them is low even without taxes (71).

Because of this tax, the development of low-grade ores, the taconites, was held up for years in Minnesota. No company could hope to mine taconite ores profitably because the cost of producing pellets from taconite is about thirty dollars per ton, and very large tonnages of reserves must be maintained to assure continued operations and to amortize the capital investment. Minnesota's taxation of reserves in the ground effectively prevented the construction of pellet plants, causing the iron companies to move their operations and develop new mines in Canada, where the tax situation was more favorable. The Minnesota iron-mining communities gradually stagnated until they were classed by the federal government as depressed areas, although they were close to large and badly needed potential reserves of iron ores.

Since 1964, when Minnesota passed the so-called taconite amendment removing ad valorem taxation for twenty-five years from the ore mined for pelletizing plants, several large mines and pellet plants have been opened, and the area now appears to be headed for some years of prosperity (23). But reserves of direct shipping ores are still subject to the ad valorem tax, an example of a state tax originated to encourage exploration, development, and extraction and continued after the changes of time caused it to have exactly the opposite effect.

The United States is one of the few nations in the world that tax a company's profits made on mining and petroleum operations abroad when they are transmitted to the home country and paid as dividends. In this country credit is allowed for foreign taxes paid, but the net result of the tax is to nullify the efforts of many governments, especially in the underdeveloped nations, to attract capital from other nations by granting tax concessions. The only way a United States company can avoid this regulation is not to return foreign-earned profits to the United States for payment as dividends, which adversely affects our balance of trade. Is private industry more effective (and less expensive to the citizen) in helping to establish industries in underdeveloped countries than is government-sponsored foreign aid?

Quotas

Quotas generally refer to imports. They are a government's official limitations on the amounts of certain commodities that may be brought into the country. Ordinarily each supplier (or supplying nation) is assigned a quantity of the commodity that it may supply without penalty, and any excess amount is subjected to a high tax. Import quotas effectively eliminate some foreign sellers from the market and are a means of perpetuating an industry's pricing policies. An import quota is a method by which a government guarantees short-run profitability to a domestic industry.

The United States, until recent shortages developed, maintained a complex system of import quotas on oil. This system was subjected, without much success before late 1973, to strong and repeated attacks, especially from the New England states, which require large amounts of fuel oil for heating. New England consumers wanted to import the then low-cost, duty-free petroleum from the Middle Eastern fields. To permit such free import of cheap fuel oil would have caused many (probably several thousand) small, marginal, and high-cost producers in the United States to close permanently, thus insuring the dependence of the country upon foreign oil. The recent history of Middle Eastern oil production shows that the producing nations have seized every opportunity to increase their oil revenues. The Congress of the United States was therefore fearful that, should the nation become dependent upon this oil, the price would quickly be raised beyond reason. What was the correct decision? Should users of petroleum have paid somewhat more than the current world market price to be assured of supplies from domestic sources, or should they have bought what was then lower-priced oil and run the risk of higher prices and shortages in the future? Another consideration in increasing imports of oil to an extent greater than absolutely necessary is the adverse effect on the already uneven balance of payments.

A quota may at times be bought and sold as a commodity. For instance, if a small refinery in Colorado has an import quota for crude oil assigned to it but has no refining facilities on the east coast, the Colorado company may sell its quota to an east coast refinery; or it may prefer to trade with a company that has surplus oil in Colorado but could use larger supplies on the east coast.

Although quotas are generally enforced by governments, sometimes producers have attempted voluntary quotas. The principal foreign exporters of iron and steel to the United States have at present voluntarily agreed to limit their exports in order to forestall the imposition of quotas by the United States government. This system seems to be working reasonably well, though small foreign producers, especially in the developing countries, appear to be taking advantage of the situation to build their industries and economies by expanding exports to the U.S.A.

A sliding scale import tax has at times been proposed in conjunction with the quota system. When the quota is exceeded, the tax is imposed; the greater the excess of imports, the greater the tax. Advocates of this system claim that it is simple, automatic, and flexible, and that it lends stability without subsidies to the minerals industries. No large government bureaus are required to administer it. The sliding scale import tax, combined with quotas, is also said to level out the frequently devastating fluctuations in the prices of mineral commodities.

Opponents of the quota system say that quotas discourage incentive and initiative and tend to keep prices higher than they would be under free competition.

Subsidies

Subsidies are government grants or gifts that support an industry. Frequently they are awarded during times of stress in order to encourage an increase in production.

Subsidies are easily abused and have not proved to be especially effective. To have any effect, they must be given equally to all producers and favoritism must be avoided.

Balance of payments

The balance of payments is the difference between a country's monetary outflow and the inflow from foreign aid, trade, and tourism—it is the difference between total outflow and total inflow of currency. When the currency inflow exceeds the outflow, a nation generally has a surplus available for trade and will prosper, at least for as long as it maintains such a balance. When the outflow exceeds the inflow over a period of years, the balance of trade is adverse, money continues to go out, the nation's economy declines, and inflation takes over.

International trade is a give-and-take affair. A nation's foreign sales and its purchases of commodities and services should nearly balance over the years so that all parties can continue trading to their mutual profit.

Duties and tariffs are not the only kind of restriction that affects international trade and the balance of payments. There are many time-consuming and unproductive rules that must be met by industries, including extractive industries, that trade internationally. For instance, the United States Department of Commerce recently issued a list of regulations which may be considered nontariff trade barriers. They apply to raw materials as well as manufactured articles. We append this list to show the amount of red tape, some of it necessary, that chokes the trade between this nation and others.

United States Government Regulations on International Trade

A. Customs law:
  1. Regulations governing the right to import.
  2. Valuation and appraisement of imported goods.
  3. Classification of goods for customs purposes.
  4. Marketing, labeling, and package requirements.
  5. Documentary requirements, including consular invoices.
  6. Measures to counteract disruptive marketing practices, e.g., antidumping and countervailing duties.
  7. Penalties—for example, fees charged for mistakes on documents.
  8. Fees assessed at customs to cover cost of handling goods.
  9. Administrative exemptions—for example, administrative authority to permit duty-free entry of goods for certain purposes.
  10. Treatment of samples and advertising materials.
  11. Prohibited and restricted imports.
  12. Administration of customs law provisions, delay in processing goods, inadequate or delayed publication of customs information.
B. Other legislation specifically applicable to imports under which restrictions are applied prior to entry of goods:
  1. Taxes.
  2. Balance of payments restrictions, including quantitative import restrictions, licensing fees, prior deposit requirements, import surcharges, credit controls on import transactions, multiple exchange rates.
  3. Restrictions imposed to protect individual industries, including measures to protect infant industries.
  4. Taxes applied to imports to compensate for indirect taxes borne by comparable domestic goods (European turnover tax).
  5. Restrictions applied for national security reasons other than customs law.
  6. State trading or the operation of enterprises granted exclusive or special import privileges.
  7. Sanitary regulations other than under customs law.
  8. Patent, trademark, and copyright regulations.
C. Other legislative and administrative trade barriers:
  1. Government purchasing regulations and practices.
  2. Domestic price control regulations.

3. Restrictions of the internal sale, distribution, and use of products.
4. Specifications, standards, and safety requirements affecting such products as electrical equipment, machinery, and automobiles.
5. Internal taxes that bear more heavily on imported goods than on domestic products—for example, automobile taxes on cars imported from Europe, based on horsepower rating.
6. Restrictions on display of goods at trade fairs and exhibitions.
7. Restrictions on advertisement of goods.

14

Politics and
the Extractive Industries II

〜〜〜〜〜〜〜〜〜〜〜〜〜〜〜〜〜〜〜〜〜〜〜〜〜〜〜〜〜〜〜

Inflation

Although inflation is an economic problem its causes are almost entirely political. Over the centuries the value of money has decreased, but so slowly that people have scarcely been aware of any change during most of the time. Only when the rate of decrease is fast and the drop in value is major do we have inflation. Inflation is an acute increase in currency and credit without a relative increase in available goods and productivity. When there is a sudden excess of money and credit without a similar increase in supplies, demands grow, prices rise, the value of the money falls. Inflation is the result of political action because government controls both currency and credit.

An unusually dramatic example was provided in the middle of the twentieth century by the president of a South American republic. This political leader, whether from filial loyalty or gratitude for support, decided to remodel his native village. Never before had its inhabitants seen such gorgeous apartments, such magnificent paseos. They even possessed a splendid lighthouse, which, as events turned out, they did not have time to operate. When the beautification was finished, it had cost the national income for one year. At this point, there was no money to pay the salaries of government workers. What could the politicos do but start the presses to print new money? With such a strong push downhill, the value of the money decreased faster and faster as the presses manufactured more and more, inflation snowballed, and the nation's economy declined rapidly. That lighthouse has never been used.

Inflation is a disturbing problem in much of the world today, including the

Fig. 14-1. Inflation: Indices of principal metal mining expenses based on the average of 1957–1959 as 100%. Source: United States Bureau of Mines.

Year	Total	Labor	Supplies	Fuel	Electrical energy
1965	102	101	103	99	101
1966	104	103	105	101	100
1967	109	111	107	104	101
1968	110	113	109	102	102
1969	114	116	113	105	103

United States, disturbing because it is responsible for recurring budget deficits and deficiencies in balance of payments that undermine the economy of any nation. The tremendous loss of purchasing power brought about by inflation is not limited to the rich investors whose savings are in bonds and fixed income securities; it influences all savings, retirement income, life insurance protection, and salary and relief checks. The price increases of goods and services affect all consumers and producers, reducing the purchasing power of their savings, salaries, and surplus for investments. A man who has retired from business finds that the amount of his income, provided by years of work and ample at the time of his retirement, grows less and less adequate to pay necessary bills; if he lives to an advanced age and inflation continues, he may face destitution. A woman who must live on an annuity discovers that it can buy fewer groceries each month or week as prices rise. Such economic maladjustments and resulting frustrations themselves contribute to the instability of a society and its government.

The standard of value of any currency is measured by its purchasing power. If the production of goods and the supply of money are kept near balance, then money saved for the future will maintain its value in the future. Inflation means that money purchases less today than the same amount purchased yesterday, and will purchase even less tomorrow.

Inflation is of great significance in all of the natural resources fields. As an example, gold mining was suspended in many places because governments kept the price of the metal fixed while inflation raised the cost of its production. While gold is really a minor commodity in terms of international trade, it well illustrates the relationship between the balance of payments and inflation. Can we in the United States afford to go abroad and try to buy the gold, the energy, and the other commodities that are needed for industry? How will we pay for these commodities? How long will they be available at prices we can afford?

Many interesting as well as frustrating problems result from inflation. For instance, our tax laws permit a company to recover the capital cost of an installation such as a mill or smelter by charging as an expense the depreciation of the original cost at a specified rate. This was designed to permit the replacement of equipment so that plants might be kept up to date. Should the depreciation allowed be the cost of the original construction or the cost of current replacement? A plant installed now would probably cost at least three times as much as the same plant installed in the middle 1940's.

Stabilization and pricing

At the beginning of a period of economic recession or slowdown, a manufacturer will first of all curtail his purchase of raw materials, and he will attempt to stretch his current inventories as far as possible. When the period of recession ends and business starts to improve, the same manufacturer will delay the renewal of purchases of raw materials as long as possible in order to assure himself that the recession is over. Then increased demand raises prices. Thus the presence or absence of an economic demand causes greater variations in production and prices in the basic minerals industries than in most others. Mining is notoriously cyclical in output, income, and prices.

Fluctuations in price of and income from mineral commodities are of great concern to many nations, especially those that depend, for their foreign exchange, upon income from these commodities. Underdeveloped countries with resources to be developed are particularly concerned with stability of markets, because without reasonable stability, effective fiscal planning is impossible for them. The need for stability similarly affects the extractive industries.

Most of the many schemes that have been proposed in attempts to stabilize the extractive industries have been relatively ineffective. Possibly because they are the most obvious, the preferred means of stabilization seem to be protective—the control of prices (through limiting production) and sales by establishing cartels and monopolies, either private or government-controlled, or both. The objectives of many of these organizations are only in part protective; basically what many of the producers are seeking is more and more revenue.

A *cartel* is defined as an association of independent undertakings in the same or similar branches of industry with a view to improving conditions of production and sale. Cartels operate through various arrangements. For convenience they are here divided into three kinds: the *association*, the *patent license agreement*, and the *combine*.

The *association* is concerned with activities that restrict production and set prices. Its members are producing companies, either in one nation or in several.

At times governments also combine to form associations, though the organizations are given other names.

The power of an association depends upon the proportion of the total output of a commodity it is able to control. A cartel, to be effective, must control about 75% of the production and marketing of a commodity, and preferably more. The greater the degree of control, the more effective the cartel. The association is also dependent upon the degree of support accorded by members to the terms of their agreement.

In certain nations, including the United States, a cartel cannot be legally enforced, and participation in one may invoke legal penalties. Because of the Sherman Anti-Trust Act, American companies have been more cautious than those of Europe in participating in cartel associations, although some United States companies have participated through export associations in accordance with the immunity provided by the Webb-Pomerene Export Trade Act. This Act states that associations organized solely for the purpose of engaging in export trade are not deemed illegal under existing anti-trust laws, provided that such associations and their agreements are not in restraint of trade in the United States or the export trade of any domestic competitor of such an association, and provided further that they are not guilty of unfair methods of competition. The enforcement of the Act falls to the Federal Trade Commission.

The *patent license agreement* is especially common among those industries where technological progress has been rapid—the manufacturing of chemical products, for example. It rests primarily upon the fact that participants hold patents and operate, or may operate, under patent licenses; the owner of the patent enjoys certain monopolistic rights in most countries. Under the patent licensing agreement each party undertakes to grant to the others a right to use the processes developed in its research, both past and future, and each concern recognizes certain territories for marketing as the private domains of the others.

The *combine* is a form of cartel that controls international markets, not by contract, but by uniting competitors under common ownership or management; the basis for market control is thus the corporate structure. In the combine's most frequent form, a small corporate entity in a particular industry is generally made part of a larger cartel pattern maintained by intercorporate contracts of a broader scope. These jointly owned subsidiaries are thus able to supplement patent and process agreements in particular markets. Sometimes international corporations, like domestic ones, are bound together by informal or indirect agreements rather than by formal ownership of stock.

Cartels are encouraged by several basic causes: maladjustment between productive facilities and market demands, excessive price fluctuations, needs of government for more revenue, and the necessity for economies of production and distribution.

Cartels are concerned essentially with regulating the output, sale, and prices of their products, the prevention of price cutting, the control of competition, the restriction of production, and other so-called "trust" goals. At the same time they do not overlook the possibilities of more efficient production through standardization, interchange of patents and processes, research, adoption of new inventions, joint purchasing, and other practices of efficient monopoly.

The results of research and patents rank among the most valuable assets of a cartel. Many companies have learned to their sorrow that the sale of patent rights to foreign companies in order to obtain some easy money frequently results in competition from abroad that they are unable to meet. Policies concerning the sale of patent rights are noticeably changing.

Defenders of the cartel system maintain that the cartel tends to stabilize production and to introduce changes gradually and carefully. They point out that cartels, like many of the largest corporations, have great amounts of capital available to them and thus are able to finance and develop mineral properties that are beyond the capabilities of smaller groups and most individuals.

Cartels as well as large corporations are more and more attempting to own their sources of raw materials. A so-called captive mine is maintained in order that the fabricator—for example, a steel company—will have a certain degree of control over the price and production of the raw material. The DeBeers Diamond Syndicate, which we have already described as one of the most effective cartels in existence, has for many years closely controlled the mining of diamonds as well as their sale. Over the years the price of diamonds has been steady, certainly in large part as a result of well directed research and superb public relations, but also because of the close control of output and marketing.

Opposition to cartels is based upon the fact that they are designed to increase the amount of money that a certain group can make at the expense of the consuming public—that the tendency is always to obtain the largest profit possible. In the United States the cartel is considered generally undesirable because it curtails competition, tends to preserve the status quo, maintains prices at high levels, and can be dangerous during times of national emergencies. The Department of Justice vigorously prosecutes any company or person appearing to act in restraint of trade or competition.

Cartels face obstacles in that they are curtailed by wars, depressions, internal friction among members, competition from nonmembers, and difficulties arising from size, distance between units, national policies, and the personalities of leaders. National animosities and jealousies also enter into the division of markets and the determination of quotas.

All of the obstacles to the formation of national combines hinder, often to an even greater degree, the formation of international combines.

The weakness or strength of a cartel stems from the commodity upon which it is built. A raw material, for example, may lose out to a synthetic one. Arsenic has been replaced by synthetic insecticides to such an extent that few young people today have ever heard of the once widely used "Paris green." It is also probably true that cartels are more effective with those commodities that are less directly essential to human living, such as diamonds.

Cartels in coal offer an especially thought-provoking study. With the exception of a few highly favored localities, coal mining during the present century has until recently been classed as a depressed industry. Many markets formerly controlled by coal were lost to oil and natural gas. We have noted (Chapter 10) that only with the greatly increased demands for cheap electric power has coal again assumed an important and profitable position.

After World War II, in an effort to achieve a degree of stability, a tight protective cartel was formed around the coal industry of the Ruhr. Later what may be classed as a government-operated cartel was formed in Great Britain. Protectionist policies were established in both places and import duties were imposed on competitive foreign coals. Serious efforts were made to improve efficiency and cut costs. Nevertheless, many of the coal mines are now financially unprofitable; they continue to operate mainly because social and economic chaos would result from their closing and because of national fear of dependence upon imported energy.

The coal mines of Great Britain furnish a good example of how political capital may be made by using certain methods of accounting to demonstrate profitability. For example, when the total cost of the production of coal from all mines is lumped together, the total operation may show a slight profit. However, when the mines are examined individually, it is found that some are highly unprofitable, being carried by a few very good properties.

The strength that the coal industry exerts over the electric power-generating industry and over the whole economy has been demonstrated in Great Britain. In February of 1972 when the coal miners struck for higher wages electric power was quickly curtailed. A compromise victory was won by the miners—at some cost to the nation as a whole. In 1973 the coal miners slowed down production, seeking a raise in wages; the government offered half of what they asked. In early 1974 the miners struck again, and the entire economy slowed drastically.

Minerals, because of their restricted distribution over the earth and the opportunities to obtain broad control of their markets, are especially susceptible to cartel organizations. Thus cartels have existed in many mineral commodities —aluminum, asbestos, bismuth, borax, coal and coke, cobalt, copper, diamonds, emeralds, iodine, iron, lead, magnesia, magnesite, marble, mercury, molybdenum, nickel, nitrates, petroleum, phosphates, potash, radium, silver, steel, sulfur, tin, titanium, tungsten, vanadium, and zinc. Most of these cartels were short-lived, but a few have persisted for many years and have effectively stabilized prices and production. Among the latter are the cartels in tin and diamonds. They have been effective because they have controlled the output from all of the principal sources of supply and because their members have consulted with consumers and have not priced their commodities out of the market. Stability at a fair price generally is desired by both producers and consumers.

Government-controlled industries ordinarily are not described as cartels, yet they operate in the same manner as do cartels, and for the same purposes. Gold is again a case in point. Control of the price of gold is in reality in the United States a tight cartel that permits a very limited free market. Here people until 1974 could not own gold except in jewelry. The restraints on marketing and the control of pricing have resulted in curtailment of production so that gold mining until recently has been almost nonexistent in areas of former prosperity.

Government-owned and -operated businesses, which are becoming increasingly common everywhere, are correctly described as both cartels and monopolies. Particularly under the communist and socialist systems no competition is permitted—government officials hold tight reins, and production is rigidly regulated.

Does the cartel hold greater promise of building a world free of economic conflicts? Is the degree of rigidity that the cartel introduces into the economic system better than the periodic fluctuations experienced in the past by the mineral markets of the world?

Plans have been proposed to stabilize both prices and production in the extractive industries by methods other than the cartel system, but they have seldom been effective and have not often even been attempted. One possible aid to stability that several mineral-producing nations are seeking is to broaden the base of their marketing. This means to sell not to one company or one nation alone, but to look for markets in as many nations as possible. Peru wants to sell its minerals, not only in the United States, but also in Japan, West Germany, and elsewhere. In this way a recession in one country will have less effect upon the producing nation.

The use of government stockpiles to control and stabilize the marketing and pricing of mineral commodities has been seriously proposed, though as a result of past experience most producers shudder at the thought of a government agency permanently in the price-fixing role. The idea is to "loan" minerals from the national stockpile during times of high demand and to have the minerals repaid or returned during slack periods.

The use of sliding scale quotas and duties to control the price of minerals and the amounts that may be imported is considered by some an effective means of maintaining a degree of stabilization. Such sliding scales are directly tied to the amounts of metal that domestic mines can produce.

Two factors described in Chapter 2 directly influence the stabilization and pricing of metals. One is the availability of scrap metals in large amounts. During times of high prices scrap is collected and sold, thus adding metals to primary supplies and exerting a stabilizing influence on the market.

The second factor, the discovery of a large new deposit, is especially applicable to those commodities whose sources of supply are few and whose total production is small. For a metal as widely produced and used as copper, it generally takes more than the discovery of a new deposit to influence either price or production to any appreciable extent. On the other hand, for a commodity like niobium, production of which is small, unit price is high, and production and demand are finely balanced, a large new discovery such as the one made recently in Brazil may completely disrupt the old trade pattern. It may also cause the price to drop and thus open new markets.

A degree of stabilization is desired by most people. The only way this has been achieved to any practical extent is through the use of the cartel system. There are reasons both for and against its use. European countries in general have decided that the advantages of the cartel system outweigh its disadvantages, while the United States holds the opposite view and prohibits by law the establishment of cartels, trusts, and monopolies.

National policy in the United States, as emphasized by Evan Just (34), contains some interesting anomalies. We preach free competition, yet tariffs and other self-sufficiency devices oppose free competition, and as government trends more toward the welfare state and socialism, free competition is pushed further into the background. Patents are licenses to monopolize for a certain period of years. To a degree the government, when it fixed the price of gold and restricted ownership, was acting in violation of its own laws. The United States government still "fixes" the price and production of natural gas, and enacts and enforces "fair trade" laws.

The question of the cartel system is an old one. Can a few large, closely controlled organizations be of greater benefit to more people than many smaller organizations? Is the largest steel company in the United States more effective as a single large company than it would be as several small competitive units?

Stockpiling

The government of the United States maintains a national stockpile of raw materials that it considers may in the future become of either strategic or critical value. The primary purpose is to have materials available in case of a national crisis such as war; the country twice has had to pay exorbitant prices for such materials in emergencies.

Materials may be removed from the national stockpile only with the approval of Congress. The contents are reviewed periodically and any surplus is sold.

Efforts have been made to use the stockpile for other purposes. For example, during the Johnson administration, so-called "surplus" materials were sold and money received was applied to help balance the budget. During March of 1973 the Nixon administration announced that it proposed to sell more than a billion dollars' worth of metals from the national stockpile in order to maintain the price of metals at a lower level than prevailed on the world market. Just the proposal to release this much metal caused the prices to drop. On April 20, 1973, the administration proposed to sell about six billion dollars' worth of stockpiled materials, an amount that would reduce the stockpile to a status of unimportance.

For the sake of stability of the markets the stockpile must be kept as free of politics as possible.

Politics and some national economies

Consider next the results of wise and unwise political actions in a few countries selected as examples—Ireland, Canada, South Africa, Australia, Chile, and Libya.

Ireland is a small nation that has taken positive and successful steps to improve its economic condition. Until the late 1950's Ireland had the lowest standard of living in northern Europe and emigration drained off most of its youth.

In 1958 Ireland drafted legislation designed to change this situation. The direct objectives were to stimulate all segments of the economy, to develop foreign trade, and to encourage the investment of foreign capital.

In order to obtain its objectives, Ireland offered government grants, training for workers, duty waivers, tax exemptions, land tenure, prospecting licenses, and state mining leases. Of particular interest to mining industrialists were the tax concessions. New enterprises were completely exempt from income taxes and corporate profit taxes for a period of four years from the date on which production commenced, and for a further four-year period one half of the profits were similarly exempt. Special allowances were also provided for capital expenditures on mine developments and for depreciation of mining equipment. In addition the government gave technical assistance that lowered the costs of prospecting and of mineral development. These aids were supplemented by advice from the well equipped, competent Irish Geological Survey (*46*).

When it became widely known that the participation of foreign companies in all aspects of mining would be welcome, and since the government of Ireland has been reasonably stable, with a record of impartiality in its treatment of foreign capital, investors did not hesitate to take advantage of government offers. Results have been far beyond expectations; several excellent mines have been developed and many jobs have been created (35).

Is Ireland better or worse for its efforts to attract foreign capital?

Another country that decided many years ago to encourage development of its mineral resources and planned ways of attracting foreign capital is Canada. Mining in Canada was nearly dormant when legislation was enacted granting tax exemptions for thirty-six months commencing with the day when a mine came into production in reasonable commercial quantities. Legislation provided also for security of land tenure and for clearly defined, impartially implemented regulations, and included recognition of the high element of risk in exploration and development and the exhaustibility of mineral deposits (14, 26).

Such favorable legislation is generally conceded to have sparked the tremendous mining developments of recent years that have made Canada one of the leading mineral producers in the world. In fact, so much foreign capital has been invested in Canada that many Canadians have voiced the fear that their country is in danger of being economically dominated by foreign interests, especially those of the United States. A number of recent legislative proposals are designed to cut back the percentage of foreign capital permitted in the minerals industries, but so far nothing of great significance seems to have been passed.

Canada has a reasonably high standard of living (GNP-C, $2,650); its economy is dependent upon the sale of raw materials. Since, in order to buy the products it needs to maintain its standards, Canada must continue to sell its raw materials in large amounts, the nation is under constant pressure to maintain exploration and production at high levels. For its continued development, Canada must obtain foreign capital, because the capital is unavailable locally.

While no very restrictive legislation has yet been passed, the simple fact that Canada has made efforts to reduce the period of tax exemption, to cut back on the depletion and depreciation allowances, and otherwise to restrict benefits accruing to investors in mining has been enough to cause risk capital to hesitate, and some is being diverted to other countries.

What position should Canada take? Clearly no nation can survive if it permits its economy to be controlled by outside sources, and ownership of resources implies a degree of control. Can a distinction between control and cooperation be drawn to mean mutual dependency?

South Africa has also encouraged mineral development over a period of many years, and has successfully attracted considerable foreign capital in spite of general opposition throughout the world to the policy of apartheid practiced in the country. South Africa provides that no government lease payments or tax will be payable until the aggregate of working profits has equalled the capital expenditures incurred during exploration and development.

Possibly one of the greatest influxes of foreign capital in the world in the raw materials fields, over a short period of time, has taken place in Australia since 1960, and foreign money until recently continued to flow into that country. Spectacular discoveries of iron, nickel, manganese, and aluminum ores, expanded development of the rutile and zircon beach sands of New South Wales, and renewed interest in tin, uranium, and copper deposits have all contributed to the unprecedented surge. A successful search for oil and gas has been conducted, especially in the Bass Straits between Australia and Tasmania, and the excellent coal fields of New South Wales and Queensland have been partly explored and developed with the expectation of supplying coking coals to Japanese steel makers. Thus laws designed to encourage investments have succeeded far beyond anybody's expectations and Australia is now one of the leading raw materials suppliers of the world. Developments have been so rapid and extensive that in the past few years many Australians are beginning to react as have the Canadians. They are seeking means to retain control of higher percentages of their raw materials industries and want more of the top positions to be held by Australian nationals.

One of the greatest dangers to venture capital, a danger that seems to be growing, is that of expropriation without adequate compensation. This is especially acute in the underdeveloped nations where governments are unstable and where serious social problems and overpopulation exist. Many examples from recent years could be given, including Chile and Libya.

For many years Chile was a favorite investing ground because of its large, rich copper deposits and its political stability. Then gradually, over a period of years, the attitude of the Chilean people and their government changed: taxes on profits were increased until in 1965 the foreign-owned mines were paying in taxes between 80% and 90% of their profits. At that time Chile announced that it planned to purchase 51% of Braden Copper Company, a Chilean subsidiary of Kennecott Copper Company. Kennecott was to continue as operator of the mine. The eighty-million-dollar purchase price to be paid by the government from profits obtained from future sales, plus money from other sources, was to be used to expand production. Taxes were to be lowered and a reasonable ceiling placed on them for the future. This plan seemed to offer an acceptable solution to the problem of greater government ownership and participation, but the political far-left of Chile never agreed to it.

A somewhat different type of agreement was reached by the Chilean government with Anaconda Mining Company and Cerro Corporation, the other two large North American operators. These companies acquiesced in expanding their operations as the government requested and in return were to be granted tax relief. If the companies opened new mines, the government was to obtain a 25% interest. Toward the end of 1970 the government arranged to purchase, from the reluctant Anaconda Mining Company, all of the company's raw materials interest in Chile within a period of ten years.

Even though these agreements with the foreign copper companies were increasingly favorable to the government, the cancellation of all of them became foreseeable in 1970 when the Marxist regime of Señor Allende was elected on a platform of nationalization of many industries, including all of the country's nonrenewable resources. In January of 1971 a bill was introduced into the Chilean Senate to make the expropriation legal. It passed easily in February, and by the middle of 1972 nationalization of the extractive industries had practically been completed. The iron and copper mining industries and the remnants of the nitrate industry were all taken over. Compensation was agreed upon in the cases of the nitrate and iron mining, but was denied for the copper industry, the government claiming that in the past the companies had accrued excess profits far above the value of the mines. In vain the companies denied this charge, alleging that their activities had at all times been not only ethical but also approved by the Chilean government. At this time the companies have exhausted all legal recourse in Chile. Unless Chile finally offers some payment, about the only way for them to obtain any compensation is in the courts of foreign countries where Chilean copper is delivered, where they can claim that the Chilean copper is theirs, that it was stolen, and that payment for it should go to them.

Why did the formerly stable and thoughtful Chilean electorate support a program of expropriation? What lies behind their nationalization policy?

The growth of Chile's population is currently at the rate of about 3.1% per year, while at the same time the economy has been expanding at a lower rate, between 2.1% and 2.5%, according to figures on its GNP published by the Chilean government in 1968, when both the price of copper and the demands for it were the highest ever recorded, and when copper accounted for about 80% of Chile's export revenue. Chile must import food for its people at a cost about equal to the income from Chile's exports of copper. But the fact that growth in the national economy has been unable to keep abreast of the growth in population means that each year the average person in Chile has a few less of the material things of life. In such a situation the poorer people generally suffer most, and unfortunately Chile has many poor people. Under these circumstances, in an effort to maintain their standard of living, the poorer Chileans were willing to try nationalization of industry, socialism, communism, or nearly anything that would hold out a promise of bettering their condition. During 1972 the rate of inflation in Chile was more than 160%.

By 1973 the communist government saw to it that the people had more money. But there was less to buy. Inflation continued to grow and black markets flourished as people queued up for every item they needed.

Will the nationalization of natural resources give the poorer people of Chile the happier future for which they hope? Chileans are excellent miners, they are good administrators, and they should be able to operate the mines profitably without foreign assistance, provided that they can obtain the necessary equipment made in foreign countries and keep their management people in mining. But if compensation, up to early 1974 denied by the Chilean courts, is not made for the copper developments, then foreign investments, which are badly needed, especially venture capital for the extractive industries, will be unavailable. Chile has lost the goodwill of many of its western trading partners, which can prove costly and take years to overcome. The problems that must be faced by future Chilean politicians, diplomats, and economists as a result of nationalization are formidable.

Chile is an example of a country that has so far been unsuccessful in its desperate effort to improve or even to maintain its standard of living. The same difficulties are found in all nations where population growth exceeds economic growth. The poor nations are getting poorer.

Another type of expropriation, one even less justified than that in Chile, took place recently in Libya, where the government seized the properties of the British Petroleum Company (Chapter 9). Great Britain, which for many years maintained small armed forces in the island sheikdoms of the Persian Gulf, decided to withdraw from the area. As the British forces went out, their places in these sheikdoms were taken by Iranian soldiers. This was done to prevent fighting among various factions on the islands, to keep the Persian Gulf open to Iranian ports, and because of the possibilities of finding additional oil fields. The Libyan military government claimed that the Iranian occupation was an infringement upon the rights of its Arab neighbors and was carried out in cooperation with the British. In retaliation Libya expropriated without compensation the properties in Libya of the British Petroleum Company, 49% of which is owned by the British government.

To protect its interests, British Petroleum Company obtained court injunctions, and these have effectively blocked the sale in the western world of the expropriated oil, which has distinct physical-chemical properties that enable it to be readily recognized. Libya has now turned to Russia and announced an agreement whereby Russia will buy the expropriated Libyan oil.

The leaders of Libya are able to take this drastic step because the country has accumulated large financial reserves and is in an especially favorable position to sell its low-sulfur oil. However, conditions can change quickly and drastically, particularly if the Suez Canal is reopened and free access becomes available to Persian Gulf oil. Risk capital is no longer flowing into Libya.

The future

Because no nation has an adequate supply of all the raw materials required by its people, each nation should have a national mineral policy that provides for intelligent exploration for and development of its mineral resources (including energy) and assures its participation in international trade on a basis of reciprocity. Every nation needs to formulate a policy that accords with what it has (in nonrenewable resources, capital, skills, and energy), that will enable it to get what it must have, and that will enable it to fit into the international scene.

A sound mineral policy for the United States should be formulated by Congress, for only Congress can know and be responsive to the needs of local industries and of local workers, as well as those of the nation as a whole. It should be administered by a competent body charged with the responsibility for raw materials. A mineral policy in the United States should consider the welfare of the nation as a whole. It should provide for the special needs of the extractive industries. It should coordinate federal and state laws, including taxation, that affect minerals. It should encourage international trade in mineral commodities while protecting domestic industries. Hasn't the time come for a national mineral policy if we are ever to have one? Over twenty years have passed since President Truman sent the Paley report to Congress.

Ideally, national mineral policies would lead to an international mineral policy in which all nations would cooperate.

Such a wise political action as the establishment of a national mineral policy depends ultimately upon a nation's people. Political actions in any nation are decided by its citizens. If they are not voters, in countries that do not hold free elections they give their support willingly or unwillingly to those in power. People determine their own economic health (their GNP and GNP-C), which in any nation depends heavily upon the extractive industries and their products.

If there are enough people choosing governments that provide for adequate education, there will be more and more people sufficiently informed to take wise political action. It is difficult to see where this circle can begin except with the people who are already educated, who have acquired basic knowledge about the earth and its limited supplies.

Only with education can people understand that a reasonably high standard of living is impossible in a nation whose population increases rapidly, forever lowering the GNP-C because the GNP (whose rate of growth cannot equal that of the population) must be divided by a larger and larger figure. Only with education can people realize that minerals and usable energy sources are both essential and finite, and thus make sure that their nation takes wise political actions to protect the raw materials of the entire earth.

Fig. 14-2. United States government stockpile inventories of metals and minerals as of December 31, 1971. Source: United States Bureau of Mines.

A. Materials with Stockpile Objectives

	Unit[1]	Objectives	Inventory[2]	Disposable
Aluminum	ST	450,000	1,279,017	829,017[3]
Aluminum oxide, fused	ST	300,000	427,475	127,017
Antimony	ST	40,700	46,747	6,047
Asbestos, amosite	ST	18,400	58,659	40,259
Asbestos, chrysotile	ST	13,700	11,835	879
Bauxite, Jamaica . . .	LDT	5,000,000	8,858,881	714,000
Bauxite, Surinam	LDT	5,300,000	5,300,000	0
Bauxite, refractory	LCT	173,000	173,000	0
Beryl	ST	28,000	40,247	2,451
Bismuth	lb	2,100,000	2,335,457	235,457
Cadmium	lb	6,000,000	10,147,904	4,147,904
Chromite, chemical	SDT	250,000	570,449	320,449
Chromite, metallurgical[5]	SDT	3,086,800	5,331,462	930,589[4]
Chromite, refractory	SDT	368,000	1,176,961	777,001
Chromium, metal	ST	3,775	8,012	4,237
Cobalt	lb	38,200,000	77,345,269	39,145,269
Columbium[5]	lb	1,176,000	9,413,090	5,753,691
Copper	ST	775,000	258,688	0
Diamond dies, small	each	25,000	25,473	0
Diamond, industrial, bort	KT	23,700,000	42,611,479	18,911,479
Diamond, industrial, stone	KT	20,000,000	25,141,634	5,141,634
Fluorspar, acid grade	SDT	540,000	890,000[6]	0
Fluorspar metallurgical	SDT	850,000	411,788	0
Graphite, Ceylon	ST	5,500	5,499	0
Graphite, Malagasy	ST	18,000	28,386	10,446

[1] *KT*, carats; *lb*, pounds; *LCT*, long calcined tons; *LDT*, long dry tons; *LT*, long tons; *SDT*, short dry tons; *ST*, short tons; *TrOz*, troy ounces; *fl*, flasks

[2] Total inventory consists of stockpile and non-stockpile grades

Fig. 14-2. (Continued)

	Unit[1]	Objectives	Inventory[2]	Disposable
Graphite, other . . .	ST . .	2,800 . .	2,800 . . .	0
Iodine	lb	8,000,000	8,011,814	0
Jewel bearings	each	57,500,000	60,110,058	0
Lead	ST	530,000	1,127,440	98,973[4]
Manganese, battery, natural	SDT	135,000	308,350	173,350
Manganese, battery, synthetic	SDT . .	1,900 . .	18,520 . .	16,620
Manganese ore, chemical A	SDT	35,000	146,914	111,914
Manganese ore, chemical B	SDT	35,000	100,838	65,838
Manganese, metallurgical[5]	SDT	4,000,000	11,189,973	6,659,570
Mercury	fl	126,500	200,105	0
Mica Muscovite Blk St/better	lb . .	6,000,000 . .	14,175,156[7] . .	7,415,656
Mica, Muscovite film, 1 & 2	lb	2,000,000	1,468,980	640
Mica, Muscovite splittings	lb	19,000,000	42,310,273	21,810,273
Mica, phlogopite block	lb	150,000	153,476	136,758
Mica, phlogopite splittings	lb	950,000	4,641,805	3,691,805
Molybdenum[5]	lb . .	0 . .	42,603,508 . .	6,090,723[4]
Nickel	ST	0	38,857	0[4]
Iridium	TrOz	17,000	17,176	184
Palladium	TrOz	1,300,000	1,254,994	0
Platinum	TrOz	555,000	452,645	0
Quartz crystals . . .	lb . .	320,000 . .	4,852,953 . .	4,532,953
Rutile	SDT	100,000	56,525	0
Sapphire & ruby	KT	18,000,000	16,305,502	0
Silicon carbide, crude	ST	30,000	196,453	0[4]
Silver (fine)	TrOz	139,500,000	139,500,000	0
Talc, steatite, block & lump	ST . .	200 . .	1,204 . .	1,004
Tantalum[5]	lb	3,400,000	4,180,504	0
Thorium oxide	lb	80,000	0[8]	0

Fig. 14-2.   (Continued)

Tin	LT	232,000	250,866	18,866
Titanium sponge	ST	33,500	35,015	8,514
Tungsten[5]	lb	. . 60,000,000 .	129,141,844 .	68,886,097
Vanadium	ST	540	3,307	2,767
Zinc	ST	560,000	1,117,913	42,677[4]

B. Other Inventories (No Stockpile Objectives)

	Unit	*Inventory*	*Disposable*
Asbestos crocidolite	ST . .	31,271 . .	31,271
Bauxite	LDT	891,000	891,000
Celestite	SDT	25,849	25,849
Diamond tools	each	64,178	64,178
Kyanite-mullite	SDT	4,820	4,820
Lithium	lb	. . 12,979,105 .	12,979,105
Magnesium	ST	98,089	98,089
Mercury	fl	7,723	7,723
Mica, Muscovite, scrap, splittings	lb	1,500,000	1,500,000
Mica, Muscovite blk, St. B/lower	lb	998,953	998,953
Mica, Muscovite film, 3rd quality . .	lb	. . 444,683 . .	444,683
Rare earths	SDT	11,841	11,841
Selenium	lb	474,774	474,774
Talc, steatite ground	ST	3,900	3,900
Thorium nitrate	lb	3,661,397[8]	3,581,397
Zirconium ore baddeleyite	SDT . .	16,114 . .	16,114
Zirconium ore, zircon	SDT	1,720	1,720

[3] Committed for sale but undelivered under long-term contracts
[4] Balance of excess pending Congressional approval
[5] Includes upgraded forms, basic material equivalent
[6] Includes 350,000 SDT credited to metallurgical fluorspar
[7] Includes 759,500 lbs credited to Mica, Muscovite Film
[8] Thorium nitrate credited as 80,000 lbs Thorium oxide

15

Conservation and
the Environmental Crisis

~~~~~~

Everyone sees the renewable resources of the earth, the flora and fauna, and most people are concerned about the preservation of scenery, primitive areas, and wildlife. Management of the large public domain is a fascinating, growing science. Seldom, however, until they are faced with shortages, problems of pollution, or unsightly change, do people think of conservation as applying to the other category of the earth's natural resources, those that are nonrenewable. Not often do they realize that renewable and nonrenewable resources are interrelated when it comes to conservation.

What is conservation?

Everyone expresses approval of the principle, but rarely do two people mean quite the same thing by *conservation*. To many people the word means *preservation* of the renewable resources—how to save lovely landscapes, how to prevent destruction and further reduction of the areas relatively unspoiled by our forefathers, how to maintain clean air and water, or how to protect or restore a diminishing wildlife species. The United States National Park Service considers itself an organization of conservationists. It encourages the preservation of large areas of unique public lands and natural wonders for recreation. No removal of or damage to vegetation or natural object is permitted, no privately owned mechanical equipment or industry (other than tourist facilities) is allowed to be installed within park boundaries.

The United States Forest Service also considers itself devoted to conservation. But conservation to the Forest Service, except in its mandated administration of the recently created Wilderness areas where entry is allowed only on foot, means *multiple use* of the land. It will permit within the national forests lumbering, grazing, mining, oil field development, hunting, and the leasing of summer home sites to the public.

Thus conservation means preservation to some, but multiple use to others, or something in between to still other groups. Peter Ellis (*19*) says, "One sees in conservation what one wants to see." The subject is highly emotional; discussions about conservation are almost always lacking in objectivity. Opinions are stated with insufficient knowledge of the facts, which results in strongly biased viewpoints. Advocates of the preservationist point of view label anyone who uses the land in any way contrary to the preservationist belief as a destroyer of the wild and a killer of flora and fauna; extremists of the other side call these preservationists impractical dreamers, unfamiliar with the real world and willing to exaggerate conditions in order to create public sentiments favorable to their cause.

Like many extreme positions, that of the preservationists in some ways implies a lack of knowledge. For instance, opposition to fires in the wild, a part of the point of view of the strict conservationist, seems reasonable unless one knows that fire is a normal facet of the environment. Without fire to clear the land and provide a propitious environment, the reproduction of the California giant redwood is inhibited. The animals of the African plains could hardly maintain their life cycle without the fires that frequently sweep over grasslands and permit new grass to grow unimpeded.

The fervor of the preservationist attitude was well expressed by Ehrlich and Ehrlich (*18*) when they said, "Each gain is temporary, but each loss is permanent. Species cannot be resurrected; places cannot be restored to their primitive state. Consequently, even if the conservationists were evenly matched against the destroyers, the battle would probably remain a losing one." *

Who are these destroyers? Anyone who uses a sheet of paper, who drives an automobile, who has a telephone, a radio, a refrigerator. Anyone who owns a television set or uses artificial light. Anyone who heats a home, who applies paint, hammers a nail, or flushes a toilet. Even the staunchest of preservationists is such a destroyer. How can the preservationists expect their battle not to remain a losing one when those they are fighting include themselves? As Pogo put it, "We have met the enemy and he is us."

The problem of conservation stems directly from the pressure of growing numbers of people and the progressive reduction of amount of wild open space. This is recognized by everyone interested in maintaining an acceptable standard of living, clean air and water, a generally clean environment, and some regions of unspoiled wilderness. It is useless, however, to attempt to

---

* Ehrlich, P. R., and Ehrlich, A. H. *Population / Resources / Environment*, 1st ed. W. H. Freeman and Company, San Francisco, 1970.

divorce man and his works from the rest of the environment because man is an integral part of the ecological system; he can no longer consider it something that exists only to serve him. "We abuse land because we regard it as a commodity belonging to us," said Aldo Leopold (37). "When we see land as a community to which we belong, we may begin to use it with love and respect."

Clearly man influences and changes the environment, but so does nature; otherwise the hairy mammoth and the dinosaurs would still be with us (49). Changes are inherent in nature, a fact well expressed by DuBridge (15): "The environment of this planet earth has been under continuous change for about 4.5 billion years. The earth never was in a stable state, for it has changed and evolved continually and radically through these many eons. In fact nature seems to abhor stability and to be in love with change."

Change cannot be avoided, although much of it can be directed. It is unrealistic to expect any family willingly to forgo the modern conveniences of life. It is unreasonable to expect people not to use deposits of needed minerals such as copper when the deposits are close at hand and their products help to lighten the burden of work. Civilized people want and must have raw materials, especially energy, at moderate prices; nations have gone to the extreme of war to obtain them. For this reason, if for no other, those who advocate the preservation of large wilderness areas known to contain valuable and necessary raw materials are not going to prevail. Actually, the minerals industries use a very small percentage of the land surface, and this use is temporary. Mineral extraction does not of necessity destroy the usefulness of land or its beauty, and many wilderness areas do not contain ore deposits or oil fields.

Here we use the term *conservation* as defined by Peter Ellis in a penetrating discussion (19): "Conservation is the effort to insure to society the maximum present and future benefit from the use of natural resources." We take conservation to mean, not the indefinite preservation of numerous large areas of unusable and inaccessible wilderness, to be visited only by the hardiest hikers, but maximum use for the benefit of the most people. This is not to say that primitive wilderness areas should not be established; they should be. But not all public land should be wilderness, any more than it should all be given over to mining. Either wilderness or mineral extraction represents only one use of the land. Conservation should mean its multiple use.

Conservation of nonrenewable resources

Conservation should mean also the most enlightened use of the land for any purpose, if the maximum benefit is to be derived. In the past, some extractive corporations and some individuals have been able to realize the greatest profit and avoid possible losses by rapidly removing the higher-grade ores, thus amortizing their investment and making a quick profit. This shortsighted practice has resulted in the loss of lower-grade materials, poor recovery in oil and gas fields, erratic prices in the market, and general waste. Fortunately most companies have now abandoned it. Many oil and gas fields are now operated at rates that permit maximum recovery, and many mines recover all marginal grades of material upon which a profit can be made.

It is impossible, however, for a company to remain long in business if it extracts much profitless material, and in the United States under present conditions, especially during times of low prices, marginal as well as submarginal materials must be left in the ground. Although a few of these marginal minerals may become valuable at later dates, most are probably lost forever, as the margin of profit may never become high enough to justify the capital investment required to reopen a mine.

It might be well worth the cost if the government could offer tax concessions or other stimuli to enable operating mines to remove marginal ores, thus preventing waste and conserving and using to a maximum our national mineral resources, as Mexico did in the case of the Boleo copper mine in Santa Rosalia (Chapter 2).

Effects of the extractive industries on the environment

If land is to provide for society the maximum present and future benefit, no single use can be allowed to spoil it for other uses. Many examples can be cited of abandoned mining areas that have become unsightly, barren dumps, dusty tailing piles, heaps of rusty equipment, and dilapidated buildings. These places are ruined for recreation and esthetic enjoyment, and sometimes for wildlife. People say, bitterly, "This is what mining does to a lovely countryside."

Not all of these examples are old mines, unfortunately; a few have been abandoned recently. Also, all too often one sees ugly bulldozer cuts and unnecessary roads marring a landscape where mining is currently in progress. Legislation to prevent this type of destruction is overdue; proper regulation of industry, particularly on public lands, is a recognized and necessary function of government. If regulatory laws do not exist, they should be passed and enforced, even though such legislation may be difficult to formulate because frequently the destruction occurs on private property and takes place with the consent of the owner. Over twenty years ago Aldo Leopold pointed out a fact that is still too true: "When the private landowner is asked to perform some unprofitable act for the good of the community, he today assents only with outstretched palm" (37). Ideally man should respect his environment to such an extent that he accepts personal responsibility for its care, instead of having responsibility forced on him by law.

In recent years the extractive industries have spent many millions of dollars in attempts to maintain an acceptable environment. In order to survive, they must cooperate with reasonable conservational and environmental objectives. All but a very few know this and have accepted these objectives as their own. Now is the time for closer cooperation between conservationists and industry, cooperation which could yield results that would be approved by all except the most extreme preservationist and the most shortsighted businessman. The goals of thinking people on both sides are now really not far apart.

Situations where the extractive industries have harmed the environment are for the most part in the past. However, situations where organizations devoted to the thesis of no further mineral development on public lands have caused the delay, for a time at least, or even the prevention of the development of mineral deposits are all in the present. These include the phosphate deposits along the shores near Savannah, Georgia; the molybdenum deposits at White Cloud, Idaho; the copper deposits on Plummer Mountain near Glacier Peak, Washington; and the chromium, nickel, and copper deposits of the Stillwater Basin in Montana. There are many others. One of the most interesting is a large copper deposit in Puerto Rico, where opposition to development has been organized by the Episcopal clergy in New York City, with the theme of preserving the culture of the poor, semiprimitive people in that part of the island. If this deposit were mined with careful protection of its environment, the culture of the local people would be changed by the addition of jobs and the improvement of their standard of living. Furthermore, the dependence of the United States upon foreign supplies would be decreased, as would inflation because of the beneficial effect on the balance of payments. Some people have suggested that the problem should be solved by pretending that the deposits do not exist, a desperate form of self-deception that would accomplish as much as a frightened ostrich does by burying its head in the sand. There must be a policy. What can it be except multiple use?

Approximately one third of the land area of the United States, most of it in Alaska, is owned by the federal government. The administration of this tremendous area is under the direction of various federal bureaus, such as the Bureau of Land Management, the National Park Service, the United States Forest Service, and the Department of Defense. To a lesser extent, overlapping control is exerted by others such as Fish and Wildlife, the Bureau of Indian Affairs, the Reclamation Service, the Bureau of Mines, and the Geological Survey. Some government bureaus are able to enforce regulations, others are not. The Forest Service can control the location of access roads and the cutting of timber; the Bureau of Land Management can prevent road building.

Additional amounts of land are owned by individual states and administered by various state bureaus.

Proper land management is a complex and enormous job. The conflicting goals of the many agencies and the needs of the public must all be taken into account. Maximum use of land for the most people cannot be achieved efficiently, and probably cannot be achieved at all, while numerous agencies with overlapping jurisdictions exercise varying amounts of control.

Laws governing mineral resources on public lands

Laws governing mineral resources on public lands in the United States are highly complicated and variable. For oil, gas, and many of the nonmetallic minerals such as potash, a system of bidding and leasing is employed. For most metal deposits in solid rock, a system of mineral claims, based upon the land laws of 1872, is still in use. These mining claims have a maximum size of six hundred feet wide by fifteen hundred feet long. For *placers* (stream gravels and beach sands containing economic minerals) or gravel deposits, the claims contain up to a maximum of twenty acres (25).

Mineral land laws have been changed but little in recent years and many proposed changes have been vigorously opposed by most mining industrialists. They claim that the present system has been eminently workable, though they will admit to some abuses that need correction. Complete rewriting of the old law of 1872 has been advocated by others, who would like to extend the leasing system to cover all nonrenewable resources.

The leasing system has its drawbacks, particularly in the way it favors the larger operations and is apt to limit healthy competition. However, the leasing system is operating successfully in Australia, South Africa, Ireland, and elsewhere.

Four concepts are basic to any realistic and effective mining law (45):
• Prospecting for minerals should be encouraged by allowing individuals and companies maximum but nonexclusive access to public lands to search for ore deposits.
• A prospector or company having found evidence of the probable presence of a mineral deposit should be given exclusive exploration rights for a limited area for a limited time while he is focusing his exploratory activity on it.
• A person or corporation having discovered a valuable mineral deposit should have an exclusive right to develop and mine it, including the right to defer such development for a reasonable period of years until economic or technological conditions justify production.
• The law must provide the person or corporation with tenure for the duration of mining, on reasonable terms set in advance.

Public trusteeship

In the United States and many other countries the doctrine of public trust provides the basic legal theory for suits to enforce consideration of environmental quality. This doctrine, the origins of which are said to be based on Roman law, holds that certain common-property resources must be held in trust for the general public. Among them are usually included the air mantle, water courses, the coastal and marginal areas of oceans, bays, lakes, and unique natural wonders.

Public trusteeship is based upon three related principles (56):

• Certain resources, such as the air and sea, have such importance to the citizenry as a whole that it would be unwise to make them subject to private ownership.

• These resources are so much a part of the bounty of nature, instead of being dependent on individual enterprise, that they should be made freely available to the entire citizenry without regard to economic status.

• It is a principal purpose of government to promote the interests of the general public. Government cannot therefore redistribute public property with a change of emphasis from broad public benefit to restricted private benefit.

A point of controversy arises when conservationists, especially those tending toward preservation, try to extend the doctrine of public trusteeship into areas that were formerly excluded, into the wilderness and the desert, in an attempt to preserve these areas, intact and unused, for future generations.

Pollution of the environment

Many people believe that pollution of the environment has reached a point where public health and safety require changes in habits of living. The air must be cleaned, scrap metals must be recycled to a much greater extent than heretofore, and the methods of disposition of the unusable solid waste must be improved. As much of the remaining waste as possible must be reused. Industry must no longer be permitted to discharge noxious debris into waterways or into the air (10). These changes, which now meet with general public approval, must be made. But to implement many of them will require time and research and a great deal of money, for many of the necessary processes and techniques remain undiscovered. Yet the program must be continued as rapidly as possible. It cannot be permitted to falter.

The extractive industries have received their share of adverse criticism from many sources. Accidental oil spills and oil leaks have aroused storms of protest;

conservation groups, backed by strong public sentiment, demand and are getting better controls. Oil fields in many areas are now being attractively landscaped and are very different from the closely packed forests of derricks in older fields. Modern mines too are being made more sightly as better housekeeping is practiced. No longer can waste be dumped into water courses. Abandoned mine dumps and tailing piles are stabilized and planted with grasses or trees. Acid mine waters are neutralized, and necessary changes are being made in most smelters to avoid their ejection of sulfurous gases into the air. When mine pits, especially the coal strip workings, are mined out and deserted, they are landscaped and restored to use for recreation or agriculture. Such practices are frequently economically as well as socially profitable, and must be not only continued but also expanded to include all open pit mining.

One question that is not easily answered is: Where is the money coming from to clean up long-abandoned mineral properties?

Recycling

The recycling and reuse of scrap metals is now receiving a great deal of public attention. This interest is perhaps spurred on by the average citizen, who dislikes seeing large areas around towns and cities choked with the skeletons of old automobiles and does not understand why these relics cannot be resmelted and their metals reused. If this could be done, a smaller amount of new ore would have to be mined.

In the past, the use of scrap metal has been strictly a matter of economics; to recycle old automobile bodies has been unprofitable, and as a general rule it still is. Who is to pay for this recovery? Old automobile bodies (not including engines) are light-weight and low-grade scrap. They must be hauled to a scrap yard where they are hammered into blocks of compact metal or are shredded before they can be shipped. Minerals other than steel in the scrap keep it from meeting the specifications to qualify for various steel products. In such a case the quality can be improved by sorting, but sorting requires the use of semiskilled labor and is expensive. Only at times of exceptionally high prices for scrap is it profitable to collect and resmelt old automobile bodies. Government or public subsidy will be required to change this situation. One proposal is that a salvage tax be paid by the purchaser at the time he buys a new automobile.

The presence of the impurities found in low-grade scrap means that only small amounts of such scrap may be used with large amounts of new ore, and the use of low-grade scrap such as that of old automobiles is being further reduced, as we have seen, by the expansion of the basic oxygen process of steel-making. On the other hand, the development of a new process of shredding is encouraging the use of scrap iron.

Recently many depositories throughout the United States have been established for the receipt of old bottles and pieces of metal, especially washed and compressed tin and aluminum cans and containers. This project has enabled the recovery of a surprisingly large amount of scrap materials, since it is being actively supported by many people. It is also proving to be economically feasible and certainly reduces the amount of solid waste thrown on dumps. The program should be expanded to all towns and cities, and all recoverable and reusable materials.

Each year the amount of scrap metals thrown away is so large that efforts to recover it are well worth consideration. The discarded materials include many metals other than iron, aluminum, and tin—silverware damaged in garbage disposal units, bits of copper wire, old tools, pieces of radios and other electronic devices, and various odds and ends. How can these materials be recovered profitably and recycled? Both industry and government should study this problem, because its solution would greatly benefit the conservation effort.

Fig. 2-4 shows the amounts of primary metals and scrap—of iron, copper, lead, zinc, tin, and silver—used in the United States in the years 1960 and 1970. It is evident that large quantities of reusable scrap are still being thrown away, though the amounts recovered and recycled have been increasing in recent years.

Unfortunately, not all raw materials can be recovered as scrap. Clearly, when gasoline is burned, it cannot be reused. Similarly, when fertilizers are spread over the ground, the additive is dispersed and cannot be recollected. Silver is spread thinly over photographic films and, while much of it has been recovered from commercial films, it cannot so far be economically recovered from the many bits of film used by individuals.

In any expansion of recycling activities we must bear one fact in mind. The cost of the energy expended is a major item and usually higher than the cost of the original extraction.

Smelting

Smelters have long been the target of conservationists because of their emission of sulfurous gases into the atmosphere and the resulting damage to vegetation in the vicinity. Considerable headway has been made in the control of the gases and the most up-to-date smelters in the United States meet the regulations established by the Environmental Protection Agency. However, several states have established their own air pollution standards that are much stricter than those set by the federal government. No smelter is able to meet some of these standards and the result is a rash of lawsuits and the curtailment of operations or closing of smelters.

A great deal of experimental work is being done in hydrometallurgy in an effort to recover metals from their ores by chemical means without the necessity of smelting. Many processes have been suggested, and the chances of success in many cases seem good.

During the smelting of iron and steel, as we have noted, large amounts of valuable minerals, such as manganese and fluorite, are used and then dispersed and lost in the slags. In spite of efforts to recover the manganese, so far they have not resulted in a process that is economic. Continued research on the recovery of materials from slags is needed.

Energy and pollution

Of all the pollutants of the environment, the mineral fuels have been the most severely attacked. The large and growing numbers of automobiles being used and the adverse publicity concerning exhaust emissions have led to widespread attempts to curb the output of harmful pollutants in gasoline. New types of motors are being tested and efforts are being made for more efficient use of the fuels in the internal combustion engine. Stringent laws have set legal time limits for the meeting of emission standards.

Coal, which is increasingly used as a source of electric power, is attacked on two counts. The strip mining practices which are so essential to keep costs low leave large areas of land as barren waste unless the abandoned pits are graded and reclaimed. All abandoned pits, whether from coal or other mining, should be reclaimed and planted.

The second criticism is directed at the smoke poured out by many coal-burning power plants. Much of the coal is high in sulfur, which will destroy surrounding vegetation unless most of it is removed before burning or from the stacks. The emission of large quantities of particulate matter results primarily from faulty design of the plant, which should be corrected.

Uranium is the third mineral fuel that has been the subject of public displeasure. As we have already noted, the radioactive waste liquors must be disposed of or stabilized in some manner, but how and where? The answer to this problem is still being sought. During the development of power in a nuclear reactor, the large amounts of hot water produced must be either cooled or allowed to run into the drainage system, where they raise the temperature of the stream. The use of cooling towers is highly recommended; they are coming into greater use. If nuclear plants maintain their near-perfect safety record, people may stop fearing leakage of radiation and be willing to have reactors built near their homes.

Mine wastes

Dumps of waste rock, open pits, and holes cannot be avoided in mining, but their impact on the environment can be controlled to a great extent. In most underground mines the openings made during mining are backfilled or refilled with waste materials. This practice prevents slumping or settling of the surface and also cuts down on the amount of waste rock that must be stored on the surface. Since broken rock occupies a larger volume than does unbroken rock, a surplus of material is always left to be stored. This rock waste cannot be permitted to enter stream drainage or contaminate farm lands. Similarly, open pit or strip mining requires areas for the storage of waste rock removed in order to get to the valuable mineral below. It is obviously impossible to restore the surface of the land in large open pits to the profile that existed before mining began. Here, however, the land can be reclaimed and landscaped to make attractive lakes surrounded by timber and grasslands, recreation areas, and even farms.

In some mines large quantities of water are encountered and must be removed for mining to begin or be maintained. In one underground iron mine in Michigan, as much as ten tons of water were removed for each ton of ore hoisted. Ordinarily this water is clean and can be put directly into streams without damage, but in a few cases it must be treated to remove excess acid. Mine waters from long-abandoned properties are particularly troublesome as at times they are highly acidic.

One of the perplexing problems facing mining operators is where to put waste rock and mill tailing. The storage place of this waste material must be where it can be retained without danger of leakage and with a minimum of disturbance of the environment. It must also be moved and stored at a reasonable cost.

An example of this kind of problem is at the property of the Reserve Mining Company on the west shore of Lake Superior. Here the company has an iron ore mine and a taconite recovery plant. It has been dumping tailing and waste rock in a deep trough in the lake after considerable study resulted in the decision, approved by a state agency, that to dump the waste there would cause the least damage to the environment. Circulation in the deep trough was determined to be at a minimum, harm to the ecology would be slight, the tailing would be confined to a very small area and would not spread over the bottom of the lake. Still, the Environmental Protection Agency has not liked the idea of dumping tailing anywhere in the lake, and, urged on by conservation groups, is attempting to force the company to seek other disposal areas.

The future

If one could foretell the future status of conservation of the environment, one would know the future of the human race. What we have to conserve is not one but all of our requirements of space, energy, and minerals (*11*). The environment we must protect is the entire earth, including its air and water. The awesome aspect of the situation is its magnitude, the interrelationship of its components, and the fact that it demands from man almost immediate action, objectivity, knowledge, wisdom, and cooperation. The hopeful aspect is that man has shown forethought about his future and is already trying to do something about the conservation of his environment.

16

"The Enemy... Is Us"*

Many years ago pastoral man learned, probably the hard way, with loss of his animals and privation if not starvation for himself, that a grazing range will support only so many sheep or cattle without being destroyed; he learned that he could not permit overgrazing if he were to prosper. Can modern man apply the same principle to people? Our range, although wider and deeper, has its limits like any other, and we are reaching the point of overgrazing it.

For the next twenty years or so, while today's children will be growing up and establishing families, the projections of population growth and demands on raw materials can be plotted with a fair degree of accuracy. These twenty years will determine the future. These twenty years are about all the time man has in which to plan and set a course. At the end of this time, if they have not done so earlier, the governments and people of the world will have to face the reality that the earth is a finite body and can support only so many people at acceptable standards of living. We will have protected our grazing range by limiting our numbers or we will have destroyed it.

Provided that man can control his population, his situation is not hopeless. He can do many things to conserve raw materials and to find additional supplies. He can improve the recovery and use of scrap metals. The almost spontaneous recent program of collecting old cans and bottles for reuse has been a long step forward, not only utilizing materials that were formerly thrown away but also making each of us conscious of our need for recycling and reducing the amount of trash scattered around the countryside.

He can force his governments to provide some form of tax relief or other incentive to encourage recovery of marginal and submarginal ores left in many

* "We have met the enemy and he is us."—Pogo.

underground mines, or oil and gas left in pools, because they cannot be extracted at a profit. When the costs in taxes and energy are too high, these resources are abandoned and lost forever.

He can work toward the development of more substitutes and make better use of those that exist. Iron and aluminum, for instance, being among the most abundant materials we possess, should whenever possible take the place of copper, which is much scarcer.

He can encourage and support technology in every aspect—methods of extraction, design, and metallurgical processes—especially in ocean mining. He can also endeavor to solve the other problems, legal and political, of ocean mining.

He can develop better exploration practices in all raw materials fields.

He can do away with planned obsolescence. It is ridiculous that people must replace household appliances like refrigerators and washing machines every few years when they can be made to last a lifetime. Automobiles can be built to last much longer than they do.

Man's most pressing problem is the energy which will be required by all of the foregoing actions. Energy resources can be used far more intelligently; who in an industrialized nation has not realized this fact while he watched rush hour traffic on the arterial highways of any large city and noted the great number of automobiles carrying one or two passengers? Urban mass transportation must come. So must more efficient motors. And we must utilize materials that require less energy for their fabrication. The prices of iron and aluminum, for instance, are about the same only because the "cheap" power needed to make aluminum has been largely subsidized by the public, as in using public funds to build dams and hydroelectric plants.

But no foreseeable conservation appears adequate. Man must find additional mineral supplies and develop new sources of energy.

Working in all these ways to provide himself with the pollution-free space, the minerals, and the energy he needs, man will constantly have to fight the difficulties caused by the interrelationships of his three needs. For instance, to avoid the pollution of air by exhaust fumes from gasoline internal combustion engines, the United States now requires pollution control devices on the exhausts of automobiles. Largely effective in achieving their purpose, these devices reduce gasoline mileage from twenty or fifteen to about ten or less per gallon. This results in an increased demand for gasoline, reflected in growing shortages and the need for additional refinery capacities. Yet, for the sake of the environment, there is opposition to the construction of new refineries, especially in the crowded urban centers where they are most needed. The refineries must be built in more isolated areas at considerable distances from the places where the gasoline is used. This means that costs and therefore prices

Sydney J. Harris

There's Just One Day Left

I read a frightening riddle the other day. It was propounded by Dr. Peter E. Glaser at a scientific conference on Energy and Humanity in London last fall. It goes like this:

A farmer has a pond with a water lily in it. The lily is doubling in size every day. In 30 days it will cover the entire pond, killing all the creatures living in it.

The farmer doesn't want this to happen, but he is busy with other chores and decides to postpone cutting back the plant until it covers half the pond.

The riddle is: on what day will the lily cover half· the pond? And the answer is: on the 29th day — leaving the farmer just one day to save his pond!

★　　★　　★

THIS IS WHAT is meant by the chilly mathematical phrase "exponential growth." Our technological society is growing at an exponential rate — and along with it our need for energy to feed this growth. We are nearing the 29th day, when we must either find new sources of energy that will not threaten our planet, or cut back drastically in our rate of production, population, and consumption.

Unless we — by which I mean the whole world — can obtain enough solar energy to replace our dwindling stock of non-renewable resources, we will continue not only to deplete these resources at an alarming rate, but also to throw the whole system of nature drastically out of kilter.

We may not yet be at the 29th day, but it is approaching far faster than the layman imagines. And if we are to deflect this catastrophe in time, it is imperative that the nations of the world unite for a common purpose: to protect the survival of this closed planetary system of the earth, and to assure that everyone will have at least enough.

★　　★　　★

INSTEAD, as Arnold Toynbee observed not long ago, since the end of World War II, there are twice as many nations and sovereignties with half as much space as before. Nationalism and separatism have proliferated everywhere on the globe; the great powers may find a modus vivendi, but the smaller nations will soon possess the capacity to make nuclear bombs and enter the arena of political and military combat.

Just at the time when we should be subordinating our provincial differences to our common global plight, we are drawing away from each other, in smaller and more fiercely tribal units. Just when we need to pool our resources for the salvation of the species, we seem bent on seizing more of the spoils. It is madness.

From STRICTLY PERSONAL by Sydney J. Harris, courtesy of Publishers-Hall Syndicate

increase as a result of transportation charges (*61*), and in most cases the additional transportation causes additional pollution of the air.

Several of the strongest conservation groups in the United States oppose the building of any new refineries, smelters, power plants, and especially nuclear power plants, and the establishment of any new oil fields and mines (*61*).

In addition to the complications caused by the interrelationships of his three needs, man must fight the problems caused by the effects on economics and politics of his requirements of space, minerals, and energy. Man will never use all the resources of the earth. They will not become completely exhausted, but as their supplies become inadequate, they will increase in price. For instance, there will always be iron. Anyone can find iron in almost any soil, as can be proved by passing a magnet over dry, loose soil in a back yard or field. Yet the cost of energy expended for the amount obtained makes this iron uneconomic. How much can we afford to pay for a commodity as it becomes more difficult to obtain, as mines and oil wells become deeper and the grade of ore available to us becomes lower? Will those nations that can pay higher prices get the commodity while poorer nations go without it? If sufficient materials are not made available at reasonable cost to all humanity, the resulting unevenness in standards of living will have international repercussions on economics and politics and will ultimately affect the quality of living everywhere.

Man is not yet united in dealing with his problems. The industrialized nations try to obtain supplies of energy and other raw materials at the lowest possible prices. For their part, the underdeveloped countries charge all they can for their raw materials because they need the income to improve their standards of living. International relationships being what they are, all nations attempt to deal from positions of strength.

Each nation, none of which is self-sufficient in its supplies of essential raw materials, struggles as an entity to obtain what it needs. The United States, for example, must consider how dependent it should permit itself to become on foreign supplies of iron ore and crude petroleum, each of which at present account for about one third of what the country uses. It must also make important decisions now that will affect the availability of energy in the nation during the next two decades (61). What is to be done in the interest of national security to strengthen the domestic production of oil and gas, considering the international implications of the nation's increasing dependence on foreign oil? What incentives are to promote the huge research effort that will be necessary if unconventional fuels are to be developed quickly? What standards to safeguard the environment are to be established that industry can achieve and consumers can afford?

Man's basic requirements of the earth transcend national boundaries and to obtain them, especially as it grows more difficult, may necessitate national mineral policies so cooperative that they constitute an international policy on minerals and energy.

Facing these problems, what is the individual to do? He must be informed about the resources of the finite earth, so that he can vote intelligently and make sure that his elected officials act intelligently. He must be aware of all sides of the problems and all their implications. He must be able to refute the statements, based on lack of knowledge, that he will unfortunately hear all around him, such as these:

• Worry about sufficient supplies of raw materials is like worry in the past that there would not be stable space for enough horses to provide transportation in New York City.

• We don't have to be concerned about mineral supplies because technology will always find a substitute when it is needed.

• Pretty soon everything can be made of plastic.

• Why worry about energy? We can always go back to windmills.

The educated individual will not be able to resort to the escapism of a point of view that we have not heretofore mentioned, which sees man's situation in a way that would provoke more or less this reaction to what we have been saying: "You have talked about supplies of energy and minerals as necessary to maintain our industrialized civilization, but our civilization is a material thing. We would be better off to let it collapse and to return to a simple life."

That simple life would be one of work from dawn to dark. There are not enough forests to provide even elementary homes. If there were, the manual energy consumed in cutting wood for homes and fires, let alone obtaining food, would give no time for thought, for recreation, for observing nature, for absorbing the literature and art and science bequeathed us by the past, or for any creative activity whatever. It is our civilization, developed by man's efforts through centuries, that has given us time for these activities and freedom to enjoy them.

The alternative to man's intelligent action in solving the problems of obtaining what he needs from the finite earth will be his retrogression, culturally and biologically, as a species. The difficulty lies not in making the choice but in recognizing its existence while there is still time to choose. Why have we not yet recognized it? Who has prevented us? Who is driving man toward such a dreadful alternative? It is time to meet the enemy, and to do so we need only look in a mirror.

Can we not work harder to control pollution and to conserve our environment and our present supplies of minerals and energy? Is it not just as necessary to develop nuclear reactors and to try to obtain energy from the burning of hydrogen? Is it not essential to limit the world's population? Can this be done in one country and not another? Can population be limited in a country with low levels of education and communication as it may more easily be in a country with high levels of education and communication? Can there be a high standard of living for all peoples regardless of numbers? Must there be a minimum standard of living for every nation? If there is to be a uniform worldwide standard of living, must there be even greater exchanges of energy and minerals? If so, must there be free international trade? What then of the necessity for world peace?

The questions are challenging (*20, 22, 48, 50, 65*). Still, unless man continues to expand his numbers, the problems are far from hopeless. If we are our own enemy, we are also the home guard. With increasing education and awareness, with concern for each other in dealing with the demands we impose upon a common habitat, can we see to it that our ignorance and selfishness are defeated by our intelligence? If we can, man will be able to maintain and perhaps to improve the quality of his life in his present home. What other chance is there for him? He has no place to go, no home but the earth.

Bibliography

1. American Iron and Steel Institute. "Steel Imports—A National Concern." July, 1970.
2. American Iron and Steel Institute. "The Steel Industry Today." *Steel Facts*. Summer, 1971, p. 12.
3. *Atlas*. World Press Company, New York. Publication discontinued.
4. *Atlas*. "Shipping Sans Suez." *Far Eastern Economic Review*, Hong Kong. October, 1971, pp. 48–51.
5. Ayers, Marshall G. "The Role of Southeast Asia in World Hydrocarbon Supply and Demand." Lecture given at Stanford University. December 6, 1971.
6. Bank of London and South America. *Review* (published monthly). Lloyds and Bolsa International Bank, Ltd.
7. Beall, J. V. "Muddling through the Energy Crisis." *Mining Engineering*. October, 1972, pp. 41–48.
8. Briscoe, M. M., President of Standard Oil Company of New Jersey. Talk before the Economic Club of Oklahoma. 1970.
9. Chase Manhattan Bank. *Outlook for Energy in the United States to 1985*. 1972. 55 pp.
10. Commoner, Barry; Carr, Michael; and Stamler, P. J. "The Causes of Pollution." *Environment*, vol. 13, pp. 2–19. 1971.
11. Creole Petroleum Company. "La Calidad Humana y el Futuro de América." *El Farol*, Venezuela. No. 233, April/May/June, 1970, pp. 20–25.
12. Dole, H. M. "America's Energy Needs and Resources." Lecture given at Stanford University. January, 1971.
13. Dole, H. M. Statement before the Joint Committee of the Congress on Defense Production, Washington, D.C. August 2, 1971.
14. Drolet, Jean-Paul. "Foreign Investment in Canadian Mining." *Mining Congress Journal*. June, 1970, pp. 38–44.

15. DuBridge, L. A. "Toward a Better Environment." *Petroleum and the Environment.* Union Oil Company. 1970.

16. DuCane, J. P. "Outlook for Mining Investments in Africa." *Mining Congress Journal.* May, 1970, pp. 60–66.

17. Durand, J. D. "The Modern Expansion of World Population." American Philosophical Society, *Proceedings,* vol. III, p. 137. June, 1967.

18. Ehrlich, P. R., and Ehrlich, A. H. *Population/Resources/Environment.* 1st ed. W. H. Freeman and Company, San Francisco. 1970. 383 pp.

19. Ellis, Peter. "Conservation—A Geological Rationale." Queensland (Australia) Government *Mining Journal.* May, 1970, pp. 3–11.

20. Faltermayer, Edmund. "Metals: The Warning Signals Are Up." *Fortune.* October, 1972, pp. 109–112, 164, 169–179.

21. Firestone, R. C. "A Road to Economic Survival." Talk given at the National Foreign Trade Council, New York. November 15, 1971.

22. Forrester, J. W. "World Dynamics: Doomsday Prophet or Man with a Message." *Christian Science Monitor.* August 7, 1971.

23. France, A. E. "Minnesota Iron Ore Taxation." *Skillings Mining Review.* 1963, pp. 10–13, 33, 35–36, 44.

24. Gaitskell, Arthur. "Risk Capital in African Countries." *Mining Journal,* London, vol. 263, pp. 237–238. 1964. Quoted from an article in *Optima,* quarterly review. Anglo-American Company.

25. Gates, P. W., and Swenson, R. W. *History of Public Land Law Development.* Written for the Public Land Law Commission. Government Printing Office, Washington, D.C. 1968. 828 pp. See especially p. 72.

26. Gibson, J. D. "Mining Tax Incentives Are Good for Canada." *Mining Engineering.* November, 1970, pp. 69–70.

27. Gill, Tom. *Land Hunger in Mexico.* Charles Lathrop Pack Forestry Foundation, Washington, D.C. 1951.

28. Hardenberg, H. J., and Reed, R. C. *General Statistics Covering Production of Michigan Iron Mines, 1971.* Michigan Geological Survey Division, Department of Natural Resources. 1972.

29. Hardin, Garrett. "Interstellar Migration and the Population Problem." *Journal of Heredity,* vol. 50, pp. 68–79. 1959.

30. Hazelett, J. M. "Present State of Arizona's Mine Taxation." University of Arizona Symposium on Mine Taxation. March 12–13, 1969, pp. 5–1 to 5–9.

31. Hendricks, T. A. *Resources of Oil, Gas, and Natural-gas Liquids in the United States and the World.* United States Geological Survey, Circular 522. 1965.

32. Hubbert, M. K. "Energy Resources." In *Resources and Man,* National Academy of Sciences, W. H. Freeman and Company, San Francisco. pp. 157–242. 1969.

33. Hubbert, M. K. "The Energy Resources of the Earth." In *Energy and Power.* W. H. Freeman and Company, San Francisco. 1971.

34. Just, Evan. "Mineral Depletion and Metal Supply." *The Technology Review.* 1950, pp. 158–159, 170–171.

35. Keist, A. J. "What the Mineral Industry Means to Australia." Seminar,

Growth: Review of Australia Economic Developments. The Committee for Economic Development of Australia. July, 1966, p. 18.

36. Lacy, W. C. "Taxation, Assessments and Ore Deposits." University of Arizona Symposium on Mine Taxation. March 12–13, 1969, pp. 2–1 to 2–25.

37. Leopold, Aldo. *A Sand County Almanac.* Oxford University Press, New York. 1949; sixth printing, 1962. 226 pp.

38. Levorsen, A. I. *Geology of Petroleum.* W. H. Freeman and Company, San Francisco. 1958. 703 pp.

39. Manners, Gerald. *The Changing World Market for Iron Ore 1950–1980.* The Johns Hopkins Press, Baltimore. 1971.

40. McGannon, H. E., Editor. *The Making, Shaping and Treating of Steel.* United States Steel Corporation, Pittsburgh. December, 1970.

41. McKelvey, V. E., and Duncan, D. C. "United States and World Resources of Energy." Symposium of the American Chemical Society on Fuel and Energy Economics, 9(2). 1965, pp. 1–17.

42. *Metals Week.* March 8, 1971.

43. *Metals Week.* March 29, 1971. "Rhodesia and the Leaky Embargo."

44. *Minerals Yearbook.* Published annually by the United States Bureau of Mines.

45. Munroe, George. "Testimony before the United States Senate Interior Committee to Change the Mining Law of 1872." *Mining Congress Journal.* October, 1971, pp. 58–60.

46. O'Neil, T. J. "The Irish Mining Renaissance." *Mining Engineering.* September, 1970, pp. 67–70.

47. Park, C. F., Jr. "Exploration and Taxation." University of Arizona Symposium on Mine Taxation. March 12–13, 1969, pp. 8–1 to 8–6.

48. Park, C. F., Jr., and Freeman, M. C. "Famine in Raw Materials by 2000 A.D." *Science Journal,* London. May, 1970, pp. 69–73.

49. Pecora, W. T. "What in the World Is Pollution?" Reprinted in *Industrial Week,* August 17, 1970, 8 pp.

50. Place, J. B. M. "Wanted: A Changed Environment for the Mining Industry." *Mining Congress Journal.* November, 1972, pp. 63–67.

51. Potential Gas Committee. "Potential Supply of Natural Gas in the United States as of December 31, 1966." Potential Gas Agency, Mineral Resources Institute, Colorado School of Mines, Golden.

52. Putnam, P. C. *Energy in the Future.* D. Van Nostrand Co., New York. 1953, pp. 16–17.

53. *Resources, Resources for the Future.* "Behind the Energy Crisis." January, 1971, pp. 1–4.

54. *Saudi Arabia Today,* vol. 9, no. 3. Embassy of Saudi Arabia, Washington, D.C. "Teheran Afterward: There's More to Oil than Money." 1971.

55. *Saudi Arabia Today.* Embassy of Saudi Arabia, Washington, D.C. "Completion of Jeddah Desalting Plant Marks Another Great Water Project." November/December, 1971.

56. Sax, Joseph. "Courts and the Environment." 1971. Quoted by *Resources, Resources for the Future* from *Saturday Review.* October 3, 1970.

57. Seaborg, G. T. "The Role of Energy." *South African Mining and Engineering Journal.* 1966, pp. 844–846, 848, 850, 852.
58. *Seventy Six.* Published bimonthly by the Union Oil Company.
59. *Seventy Six.* Union Oil Company. "The Trans-Alaska Pipeline." July./August, 1971, pp. 16–21.
60. Shell Oil Company. *The National Energy Problem.* 1972, p. 20.
61. Shell Oil Company. *The National Energy Position.* 1972, p. 19.
62. Standard Oil Company of California *Bulletin* (published quarterly).
63. *Standard Oiler.* Published monthly by Standard Oil Company of California.
64. Starr, Chauncey. "Energy and Power." In *Energy and Power.* W. H. Freeman and Company, San Francisco. 1971, pp. 3–15.
65. Sutulov, Alexander. *Minerals in World Affairs.* University of Utah Printing Services, Salt Lake City. 1972. 200 pp.
66. Tucker, J. F. "Countdown for an Arctic Odyssey." *The Humble Way,* vol. 8, no. 2, pp. 1–5. 1969.
67. United Nations. *The Future Growth of World Population.* Department of Economic and Social Affairs, New York. 1958, p. 17.
68. United States Department of the Interior. First Annual Report of the Secretary of the Interior under the Mining and Minerals Policy Act of 1970. March, 1972.
69. *U. S. News and World Report.* "Myth of Japan's 'Cheap Labor.'" January 10, 1972, pp. 48–49.
70. Waldenstrom, Erland. "The LAMCO Project." *The Annals of the Swedish Iron-masters Association,* vol. 147, pp. 437–460. 1963.
71. Weaton, G. F. "A History of Minnesota Mining as Influenced by Taxation." University of Arizona Symposium on Mine Taxation. March 12–13, 1969, pp. 7–1 to 7–30.

Index